MEN OF IRON

THE AUTHOR

SALLY DUGAN started her writing career as a newspaper journalist with the *Oxford Mail* and the *Oxford Times*. She has also written for *The Times* and the *Radio Times*. She first became interested in nineteenth-century cultural history while studying for an MA in Victorian Studies at Birkbeck College, London. *Men of Iron* is her fourth book written in collaboration with Windfall Films. She was joint author, with her husband David Dugan, of *The Day the World Took Off: The Roots of the Industrial Revolution* (2000). Her other previous books are *Commando* (2001) and *Measure for Measure* (1993).

Her home is in an Oxfordshire village within a stone's throw of a Great Western Railway station.

MEN of IRON

MEN *of* IRON

BRUNEL, STEPHENSON
and the inventions that shaped the modern world

SALLY DUGAN

PAN BOOKS

First published 2003 by Channel 4 Books

This edition published 2005 by Pan Books
an imprint of Pan Macmillan Ltd
Pan Macmillan, 20 New Wharf Road, London N1 9RR
Basingstoke and Oxford
Associated companies throughout the world
www.panmacmillan.com

0 3304 3279 6

9 8 7 6 5 4 3 2 1

A CIP catalogue record for this book is available from the
British Library.

Designed by Perfect Bound Ltd
Colour Reproduction by Aylesbury Studios Ltd
Printed by Butler and Tanner

This book was originally published to
accompany the television series *Men of Iron*
made by Windfall Films for Channel 4.
Series Producer: Ian Duncan.

Dedicated to Dr Gillian Duncan

CONTENTS

ACKNOWLEDGEMENTS

This book – and the television programmes it accompanies – would not have been possible without the aid of an army of experts. I would like to thank Anthony Burton, Ed McCann, Professor Andrew Lambert, Dr Simon Schaffer, Adrian Vaughan and Chris Wise, all contributors to the Channel 4 *Men of Iron* series.

Michael Richardson and his team of archivists at Bristol University gave patient guidance through their wonderful collection of original notebooks, letters and diaries. Michael Chrimes, Head Librarian at the Institution of Civil Engineers, kindly read and made detailed comments on the manuscript. Staff at the London Library, the Bodleian Library, Oxford, and the National Archives at Kew also offered much practical help.

At Windfall Films, Ian Duncan, Robert Hartel and Leesa Rumley provided unflagging inspiration and enthusiasm, and invaluable research and assistance came from Yvonne Bainton, Cherry Brewer, Sarah Day, Daniel Fromm, and Val Prodromou.

I am also grateful to Mandie Howard, of 4 Ventures, who got the book off the ground, to the editorial team at Macmillan – Rob Dimery, Sarah Macdonald and Emma Marriott – and the designer, Dan Newman. Finally, I would like to thank my husband, David Dugan, for his moral support and constructive criticisms.

INTRODUCTION

On 25 December 1858, two grand old men of Victorian engineering – Isambard Kingdom Brunel and Robert Stephenson – shared a Christmas dinner at the Hotel d'Orient in Cairo. Nothing so very remarkable in that, you might think. Except that these two friends, who laughed and joked over the brandy, had been lifetime commercial rivals. And before the next year was out, both would be dead.

Brunel, the French-educated Southerner and flamboyant showman, and Stephenson, the modest, cautious Northerner, could not have been more different. Brunel was full of theory; Stephenson was more practical. Brunel was a liberal; Stephenson a conservative. But both shared a passion for the unprecedented technical challenges posed by the new Railway Age.

This book tells the story of the astonishing friendship that developed between them as they each carved their names in iron across the landscape of Britain. It also looks at the secrets behind their massive constructions, which leapt over fields, hills, bogs and rivers, and still form the backbone of the nation's transport network today.

If you take a train from London to Penzance, you will cross a bridge so subtly designed that the best engineering brains of today have only just worked out how it was put together. The Royal Albert Bridge at Saltash, crossed by some thirty high-speed trains a day, is just one of any number of working memorials to Brunel's inventive genius.

Brunel was a true polymath, designing tunnels, bridges, railways, world-beating iron ships – even a prefabricated hospital for use in the Crimean War. What he lacked in experience, he made up for in confidence; he had an uncanny knack of persuading people to part with massive sums of money to fund his grandiose schemes.

He had started out working with his father, Marc, in the foetid and dangerous atmosphere of the Thames Tunnel. Through Bristol contacts made when designing the Clifton Suspension Bridge, Brunel came to mastermind one of the largest railway empires in the country. When he was first appointed as engineer to the Great Western Railway (GWR) at the age of twenty-six, he had no railway experience whatsoever. But he had taken a ride on George and Robert Stephenson's Liverpool to Manchester Railway, and was convinced he could do better.

Delegating was never his strong point. He appointed assistants to survey the line between London and Bristol, and spent his time riding round the countryside checking up on them. He commissioned steam engines to his own specifications, then had to buy some more in the light of the humiliating fact that they did not work properly. In the 1830s, the best engines in the country came from the workshops of his arch rival, the designer of the *Rocket*, Robert Stephenson.

Robert Stephenson spent his early career overshadowed by his father, a semi-literate self-made man. In fact, for many years it was George rather than Robert who was credited with the design of the *Rocket*. Robert's talents – like his father's – were essentially practical. His skill lay in adapting existing designs rather than thinking things through from first principles. He was certainly not a polymath on the Brunel scale.

The rivalry between Brunel and Stephenson became public over the so-called 'gauge wars'. Brunel had designed his railway lines to an expansive 7-ft (2.1-m) width, performing a battery of scientific tests to show that this provided a smoother ride. Stephenson had gone for a cheaper and more conservative narrow gauge. Both gave evidence to a Royal Commission set up to decide which gauge should serve the country; both had houses near to Parliament so they could lobby the decision-makers.

At the height of the Victorian railway mania, both promoted rival bids for lines to serve the same stretch of countryside. Brunel championed the cause

△ Previous page: View of the Royal Albert Bridge at Saltash. The inscription, I K BRUNEL ENGINEER 1859, was added to the portals at either end as a tribute after his death.

of the atmospheric railway, which used air, instead of a locomotive engine, to propel carriages along the track. This would have rendered Stephenson's steam locomotives redundant, and he produced a very sceptical report on the system. The report was prophetic; Brunel's flirtation with atmospheric railways in South Devon proved the most expensive failure in engineering history of his time. Yet in all this, the two men somehow managed to maintain an increasingly warm personal friendship.

Part of their secret was a refusal to make political capital out of each other's mistakes. In the early days of the GWR, a dissatisfied group of directors invited Stephenson to provide an independent engineer's report on the line; he refused. Brunel repaid the compliment at the public inquiry into the Dee Bridge disaster of 1847. Stephenson's cast-iron bridge collapsed, killing five people, and he narrowly escaped a manslaughter conviction. Brunel, reluctantly called as an expert witness, would not condemn the use of cast iron as a building material, even though he rarely used it himself.

Their mutual support extended well beyond the committee room. When Stephenson floated the giant tubes for the Conway and Britannia bridges in North Wales, Brunel was there to offer advice. When Brunel, in his turn, came close to despair during his last desperate attempts to launch the *Great Eastern*, Stephenson turned up to paddle about in the Thames mud, against doctor's orders. Stephenson's own experience of ships was limited to the luxury yacht *Titania*, which he took on cruises in a vain attempt to restore his rapidly diminishing health. However, that did not stop him from being on hand when required to lend his friend moral support.

The two men died within a month of each other, both worn out from overwork. They were not the only casualties of their ambitions. Some estimates put the death toll on just one stretch of tunnel on the GWR (nicknamed 'God's Wonderful Railway') at up to one hundred. However, it could be argued that the navvies who sweated under their orders were not asked to take risks the two men were unwilling to take themselves. Brunel, in particular, was always to be found at the centre of the action, and had several narrow escapes from death.

Creating the dramatic new landscape that was to shape the modern world exacted a price. It was a price measured in human lives, as well as pounds, shillings and pence. But, as Brunel's faithful locomotive engineer Daniel Gooch put it, 'Great things are not done by those who sit down and count the cost of every thought and act.'

THE GREAT BORE: THE THAMES TUNNEL

The romantic image of early 19th-century London, with horses clip-clopping their way through narrow alleyways, disguises a horribly familiar problem – crippling traffic congestion.

Worst affected were the busy dockyards in the east of the city, cut in half by the open sewer that was the River Thames. Foot passengers could hold their noses and queue for the ferry. Those with horses and carts had to fight through the crowds to reach London Bridge, up to 4 miles (6.5 km) away – and there was a toll to pay once they got there. One obvious solution might have been to build more bridges. However, this was easier said than done, since any bridge would have to be high enough to allow boats to pass through.

Enter a French engineer, and a worm. The French engineer was Marc Brunel, father of Isambard. The worm was *Teredo navalis*, a destructive little beast whose efficiency in boring through ships' timbers gave Brunel Senior the idea for a revolutionary method of burrowing under water. His patent tunnelling shield enabled him to create the first modern underwater transport tunnel in the world – the Thames Tunnel.

It was in the rank-smelling and hazardous environment of this tunnel, working under constant threat of being swept away by a slurry of sewage, that young Isambard served his engineering apprenticeship. Twice he narrowly escaped drowning as the river flooded into the tunnel. Every day he ran the risk of contracting the mysterious 'tunnel sickness' that lingered in the foul air, striking men blind without warning. As others fell victim to illness, Isambard found himself Resident Engineer at the age of only twenty. It was a massive vote of confidence from his father, and the best training he could possibly hope for. Provided, that is, he lived to tell the tale.

Marc Brunel was a French émigré who, though a Royalist at the time of the Revolution, had escaped the guillotine. Within a month of arriving in Britain from America in 1799, he had filed a patent for a copying machine using linked quill pens. This was followed by a system for mass-producing dockyard pulley blocks for the Royal Navy, which brought recognition but little money. Altogether he was granted eighteen patents for everything from sawmills and the manufacture of tinfoil to printing plates and marine steam engines. He was an exceptionally talented engineer, and it is typical of the unfairness of history that his son's fame should have eclipsed his.

Isambard, born in 1806, was the third child of Marc's marriage to an Englishwoman, Sophia Kingdom. From the moment he taught him to draw perfect circles freehand at the age of four, Marc was determined that his son should be a great engineer. He gave him arithmetic and geometry lessons at home, and taught him what he called 'the Engineer's Alphabet' – the habit of making scale drawings of buildings of interest. Isambard's granddaughter, Lady Noble, suggests in her somewhat romanticized account that he had mastered Euclid by the age of eight. L T C Rolt, in his biography, puts it even earlier – at six! Whether or not this is true, Isambard was undoubtedly a precocious child.

Despite being strapped for cash, Marc sent his son to an expensive boarding school in Hove, where Isambard spent his spare time carrying out a meticulous survey of the town. In 1820, at the age of fourteen – peace having broken out in Europe – Isambard went to France. Here he studied advanced mathematics at the Lycée Henri-Quatre in Paris, and worked under the famous clock-and-instrument maker Abraham Louis Breguet. Ironically – considering he was a Royalist – Marc had considered sending Isambard to the École Polytechnique, the Polytechnic School founded by Napoleon. However, although Isambard was bilingual, his English birth disqualified him from entry.

△ Previous page: Portrait of Marc Isambard Brunel, painted by Samuel Drummond around 1835, with the Thames Tunnel in the background.

▷ Marc Brunel's machinery for making ships' pulley blocks was an early example of mass-production. Manufacture had previously involved at least twenty separate operations.

Even without this finishing touch, his was an education that gave Isambard an undoubted edge over more insular contemporaries. As historian Simon Schaffer puts it:

'In the first decades of the 19th century, France was the leading scientific nation in the world. This is partly because Napoleon's regime had invested extremely heavily in major scientific teaching and research institutions. Paris had the very first teaching laboratories in the sciences, at the Polytechnic

School, and in the colleges of military engineering. Even though the French were defeated at the end of the Napoleonic War, they weren't defeated in the classroom. What Brunel was learning in Paris was state-of-the-art engineering analysis.

'The French tended to admire British achievements in practical engineering. French experts often visited Britain in the 1810s and 1820s, partly as industrial tourists and partly as commercial spies. What they learned were bits and pieces of extremely important engineering technique, for example in canal design, in the young science of railway engineering, in ship design and so on. Most of these secrets were stolen or copied down in notebooks and then smuggled out of Britain to France. But it was the French ability to combine this data into powerful models of structural stability and engineering design that made the French curricula so powerful.'

While Isambard was turning himself into a cutting-edge engineer in France, the family suffered what was ever afterwards referred to euphemistically as 'The Misfortune'.

Marc Brunel may have been a genius at the drawing board, but he was no accountant. Two circumstances conspired to land him deep in debt. The first was a devastating fire at the Battersea sawmill into which he had sunk all his savings. The second was when he was left the unwilling owner of thousands of pairs of useless, unpaid-for army boots.

The idea of going into boot production had come to him after watching troops stumbling back from the Battle of Corunna in 1812. Corunna, a Spanish village, was to the Napoleonic Wars what Dunkirk was to World War Two. British troops, under heavy fire, were forced to beat a strategic retreat by sea after a long, exhausting march. The shocking sight of the men's unshod, lacerated feet as they arrived home had prompted Marc to design a superior, machine-made boot. The government encouraged Marc to produce enough for the entire army, and so he set up a factory – manned by disabled soldiers. However, when war with France ended unexpectedly early, the government refused to pay up.

On 14 May 1821, Marc was arrested for debt and confined to King's Bench Prison in Southwark, where his wife, Sophia, accompanied him. The King's Bench was a larger version of the Marshalsea, described by Charles Dickens in *Little Dorrit*. Those with money could pay for the right to walk in the streets around the prison; those without had to stay inside. One visitor recorded finding Marc sitting at a table littered with page upon page of mathematical calculations, while Sophia sat opposite, darning stockings.

△ King's Bench Prison in Southwark, London, where Marc Brunel was imprisoned for debt in 1821 despite his services to the government.

'One of the lessons that we can learn from Marc Brunel's life is never trust the government,' says historian Adrian Vaughan.

'It is definitely arguable to say that he enabled the British to win the Napoleonic War. His wonderful machines that enabled British warships to be into the docks, repaired and out again so much quicker than they had been meant that the Navy had more ships operational.

'The army was wearing Brunel's boots at Waterloo and maybe that helped them the better to stand on blood-soaked grass when the French cavalry charged, time after time after time, and they stood their ground. So Marc Brunel did all that and owing to the reluctance of the government to pay what they owed him, plus the misfortune of the sawmill burning down, he ended up in prison.'

It was only after Marc let it be known that he was considering accepting an offer to work for the Tsar of Russia that the government was shamed into action. The Duke of Wellington stepped in to raise the £5,000 needed for his release, on condition that he remained in England.

Although Marc's stay in prison had lasted barely three months, and the Duke of Wellington's intervention was a welcome vote of confidence from his adopted country, there was no doubting the social disgrace of his incarceration. As Simon Schaffer puts it:

'Think about the problems of trust and credit in the years around 1800. In English we use exactly the same words to mean cash and to mean social status. It's a kind of besetting sin of the English. So a word like "credit" means how much money I've got in my bank balance, and how believable I am. A word like "trust" means whether I should believe what you say, but also whether I should invest in you.

'Traditionally folk who were trusted by social elites were gentlemen. You didn't trust women, foreigners, people who were sick or crazy; you didn't trust people who were doing a job for money, because they had an interest, or so it was said. So the institutionalization of the engineering profession from the late 18th century onwards means somehow turning these great engineers into men who are status-bearers, who are credit-worthy, who can be trusted. They had somehow to appear on the public stage as folk who were both experts and in a way genteel.

'Now for that project, losing all your cash, going into debt, going broke, failing, tying your tie the wrong way, not being dressed appropriately… All these could actually have rather important, if not fatal consequences for the fate of the projects in which you were engaged. So they're walking on eggshells, are these engineers, and the slightest social gaffe can often have enormous commercial and industrial implications.'

Marc was lucky. He had friends in high places, and there was a general recognition that his incarceration had been unfair. Once released from prison, he set about rebuilding his career, opening an office with a single clerk in The Poultry, Cheapside. He was joined there by the sixteen-year-old Isambard. Projects included two iron suspension bridges for the island now known as Réunion, in the Indian Ocean, a swing bridge, and a floating pier for Liverpool Docks. Other preoccupations included a series of experiments with gas engines, and ways of improving marine steam engines. This last foreshadows the obsession with ships that was to dominate Isambard's latter years.

Three years before The Misfortune, Marc had taken out a patent for an iron tunnelling shield, but – for obvious reasons – had never had a chance to put it into action. Now he was to embark on his most ambitious venture – the Thames Tunnel.

The idea of tunnelling under the River Thames was not new. Richard Trevithick, the Cornish mining engineer and inventor of the steam locomotive, had tried to make a passage the conventional way, using pick and shovel and pit props. However, the river kept breaking through the soft ground, and the project had to be abandoned when the tunnel was only 200 ft (61 m) or so from its target.

Despite this earlier failure, investors were impressed with Marc Brunel's plans. Meetings were held in taverns and coffee-houses, a company was formed, a bill passed through Parliament, and Marc was appointed engineer at £1,000 a year for three years. As things turned out, it was to be six times that long before the tunnel opened, having acquired in the meantime the uncomplimentary nickname of 'The Great Bore'.

The entire Brunel family moved to Blackfriars (not then the most select of districts) in order to be nearer the site of this new project. Borings were taken, which indicated that there was a strong layer of clay suitable for excavation, and work started at Rotherhithe on 2 March 1825. There was a brick-laying ceremony – the first of many public relations ventures – with Marc laying the first brick, and Isambard the second. Church bells rang and there were speeches and toasts, and a feast for 200 people – the table adorned by a large sugar model of the tunnel.

The sealed-up remains of Trevithick's tunnel were only about $\frac{1}{3}$ mile (1 km) away from the new workings, but Marc Brunel was confident. He hoped to avoid the problems with quicksand, which had plagued his predecessors, by placing the tunnel roof only 14 ft (4.3 m) below the riverbed. The plan was for a two-way road tunnel, with wide spiral ramps leading down from the surface.

His first move was to sink a massive shaft on dry land by the ingenious expedient of raising a 42-ft (12.8-m) high brick tower and allowing it to sink under its own weight.

Within three weeks, the tower had reached its full height, each brick-layer laying a thousand bricks a day. Celebrities – including the Dukes of Wellington, Cambridge and Northumberland and the Austrian Ambassador – joined the merely curious as they watched the progress of this mammoth undertaking.

The brickwork was strengthened with cast-iron ties, and the base of the tower was one enormous iron ring. Once the cement had set, labourers hacked away to remove soil inside this ring, in order to enable the tower to start sinking.

Marc Brunel got the idea for this technique from well-sinking technology, as engineer Ed McCann explains:

'This is a giant pastry-cutter, which sits on the surface of the ground. He builds his tower on top of it until it's very heavy, and then he nibbles away at the inside edge of the pastry cutter and the whole thing works its way down through the ground.

'Unfortunately for him, he's got to push this thing against friction, against the sides of the shaft, and he's got his sums wrong. Or maybe he didn't do any sums. Halfway down, they're nibbling away and the thing stops sinking. So he raids the local community, gets every brick in sight, stacks it up on top of the thing, and gets it moving again. Off he goes, down to the bottom.

'I worked on a project a couple of years ago where we essentially had exactly the same problem. Now we know you've got to do this quickly and in one operation. When you sink a shaft you've got to… bang, bang, bang! You keep it moving. You don't stop for a weekend, you just keep going all the way down. He may not have known that. He may have stopped on a Friday night, gone off, come back on Monday morning, started digging again and it didn't go down any further because the ground relaxed into it.'

Richard Beamish, who published the first biography of Marc Brunel in 1862, estimated that first 8,000, then 50,000 bricks had to be piled up before the shaft started sinking again. There were other practical problems, too. At first, human chain gangs had to be used to remove water and soil from the workings, since the specially ordered steam engine had not turned up. A substitute engine was found – to pump the water and to take the procession of buckets of debris to the top of the shaft – but this broke down. At one stage, too, the whole tower leaned alarmingly as it hit a patch of hard ground to one side.

However, as work progressed more smoothly complacency set in. There is a distinctly censorious tone in Beamish's account of the tunnel's first fatal accident:

So admirable had been the arrangements, and so continuous and successful the progress of these gigantic operations, that the men employed began to treat them as ordinary work, and to give way to indulgences only too common amongst their class, but from which they had hitherto been restrained by the very novelty of their engagements. The confidence, however, which had been gained proved unfortunately fatal to one of the gangers named Painter, who, coming intoxicated to the works on the night of

12 July, fell from the top of the shaft to the bottom: strange to say, he survived the concussion twelve hours.

Meanwhile Marc Brunel was perfecting the design for his tunnelling shield. He decided to make it rectangular rather than cylindrical – as in the original patent – but the principle remained the same. In his observations of *Teredo navalis* in Chatham Dockyard, he had been struck by the twin-protection system of this timber-boring worm. Digging with shells on either side of its head, the worm expels the chewed-up wood out of its body behind it. This waste then lines and reinforces the hole as it moves along. Similarly, Marc Brunel's great iron cage sheltered tunnellers, and enabled them to work back to back with bricklayers. This made a disastrous tunnel collapse less likely. At least, that was the theory.

The shield consisted of twelve cast-iron frames, six for each of the twin arches of the tunnel. It was divided into thirty-six working cells, like a giant iron honeycomb. Construction of this massive piece of ironwork was entrusted to the great engineering firm of Henry Maudslay.

As the shield was a piece of precision engineering, dependent for its safe functioning on standardized parts, Maudslay was the obvious man for the job. Known as the father of the modern machine shop, he had pioneered new standards of accuracy in machine tooling. Both Brunels made frequent visits to his works, situated in a disused riding school in Lambeth, and this contact with the greatest mechanic of the age was a vital part of Isambard's training. As Simon Schaffer explains:

'The single most important set of workshops in British engineering at this period are those run by Henry Maudslay and his allies south of the Thames in Lambeth, around the Elephant and Castle. Here you had an entire network of machine shops, not just run by Maudslay but also by people who'd worked under him. These were the engineers who not only go on to work for Brunel, but also worked for other major entrepreneurs of the period.'

Maudslay had craft skills that could not be learnt from books, and without his technical know-how – passed on by word of mouth to his apprentices – both of the Brunels' designs would have remained nothing more than marks on paper.

By the end of November 1825, the shaft was deep enough for tunnelling to begin. The great shield was lowered into place, ready to start its crawl under the river. The tunnel face was partitioned into tiny sections, each one covered by a removable board. A miner would remove only one board at a

◁ Above: Contemporary
engravings underlined
the audacity of the
Thames Tunnel works by
showing workers
beavering away beneath
a river crammed with
sailing ships.
Below: Section showing
the movable tunnel
shield, and the stage
behind for the removal
of soil.

time, dig out the small area behind it to a depth of several inches, then replace the board and do likewise with the one next to it. When a miner had excavated his part of the face, his cell would be inched forward by jacks at the rear of the shield. The bricklayers followed immediately behind, giving the tunnel a permanent lining.

It was slow work, but all seemed to be going well until, in January 1826, came the first sign of trouble: a sudden inrush of water. Marc's original plans had included a drain below the tunnel, but – alarmed at the mounting cost of the enterprise – the tunnel directors had ruled this out. This proved to be a dangerous false economy. It soon became obvious that the solid band of clay that the geologists had so confidently promised was more like a wavy line, mixed with sand and gravel. The problem was compounded by the fact that the tunnel sloped downwards. As many as forty men had to be employed using bucket pumps to clear water from the shield.

Leaking water was by no means the only problem. As engineer Ed McCann puts it:

'What you have to remember is that the River Thames at this time is a giant toilet. Basically the way people deal with rubbish, even household rubbish in London at that time, is they lob it into the street and it gets washed down into the local river, which ends up in the Thames. So the bottom maybe 10 m [33 ft] of the River Thames is compost rotting, dead fish, animal carcasses, all manner of very unpleasant stuff rotting down over many, many years, along with all sorts of mud and silt and all the rest of it.

'These guys are tunnelling away and from time to time their tunnel line comes out of this band of clay, which is quite safe, and enters into this awful gloop, which will contain – because it's rotting down – pockets of methane and hydrogen sulphide stinking of rotten eggs. I imagine it was quite scary.'

Throughout the period of the tunnel's construction, sickness claimed a much heavier toll than accident. Marc Brunel himself fell ill from pleurisy, and in April 1826, William Armstrong, the Resident Engineer, resigned his post through illness. Isambard – who had been Armstrong's assistant – was promoted in his place. As Adrian Vaughan explains, Isambard did his utmost to justify the faith his father had put in him.

'You've got Isambard spending weeks without coming up for air on the tunnelling shield, sleeping on bales of straw. This was his big one, this was where he was going to make his name. His parents became really worried about him – I mean, men were almost dropping down dead. They were certainly fainting, and he was not.

'As a result of his amazing devotion to duty, the workmen really loved him, because he led from the front. He was a great leader because he set this terrific example of working hard. Of course the other half of that is that they then joined him in his recklessness.'

Worried that Isambard was going to work himself to death, Marc found three engineers to assist him. One was Richard Beamish, a 28-year-old ex-soldier whom we have already encountered as Marc's biographer. He had resigned his commission in the Coldstream Guards to study engineering independently, and initially worked as a volunteer. The second was William Gravatt, who was only slightly younger than Isambard, and the third was a 'Mr Riley'.

Isambard's youth and his small stature – he was only 5 ft 3 in (1.6 m) – do not seem to have hindered him. This can clearly be seen in the pages of his Thames Tunnel Journal, a meticulous record of those exhausting years, now in the archives of the University of Bristol. The entries are closely written in a tiny copperplate hand, with line drawings and marginal annotations. At sensitive moments, he breaks into shorthand.

The first entry in the journal, dated Friday 29 October 1826, notes that he fined the workmen for carelessness in breaking a top plate of the shield. (The introduction of piecework had meant that the miners came under constant pressure to move fast, so the bricklayers could lay as many bricks as possible.)

That same day, not trusting the night shift, Isambard, Marc and Beamish took turns in supervising throughout the night. Isambard went to bed at 9.30 p.m., was up from 11 p.m. until 5 a.m., went to sleep again, and then was up again at 9 a.m.

Living like moles in the unhealthy atmosphere of the tunnel took its inevitable toll. One Saturday night, while making all safe for the Sunday day of rest, Beamish was suddenly struck blind in the left eye. Although he was back at work within a month, he never recovered his vision entirely.

Fever and delirium were commonplace, and often fatal. A series of journal entries for February 1827, which are the first to mention the unfortunate 24-year-old assistant, Riley, tell their own story:

Thursday 1 February 1827. Mr Riley still unwell. Wrote to his brother at his request.
Friday 2 February. Mr Riley's brother arrived.

Tuesday 6 February. Went to see Mr Riley. He is much worse. I am afraid no hopes. Dr Gaitskill says so – Mr Beamish below. A number of men ill – great delay.

Thursday 8 February. Riley much worse.

The moment of Riley's death – which took place on the Thursday – is not recorded, but the entry for Monday 12 February describes the funeral in curiously unemotional terms:

9 a.m. Started in a mourning coach with Gravatt and the two brothers. Having arrived at the Dissenter's burying ground Bunhill Row the coffin was lowered into the grave at 10 o'clock. Thus ended poor Riley's connection with us and the Tunnel.

He certainly was an amiable young man and intelligent but no energy of character and certainly not fit for our work nor likely to have become so.

There are occasional signs that the young Isambard was having difficulty asserting his authority; even at this early stage of his career, he did not enjoy being crossed. An example is the entry for Saturday 3 March:

In paying the men this day, Mr Butler again interfered in a manner which I do not choose to suffer. I had signed a bill of Jenkins for lightening barges which I had authorised him to do. On presenting it Mr Butler without giving any explanation or speaking to me (altho' close at hand) said he would not pay notwithstanding my signature was upon it. On my requesting an explanation he said he had a right to pay or not as he thought fit any bill altho' I might order the work to be done and sign the bill, and that moreover he should always represent to the committee any expenditure which appeared to him unnecessary, as a justification of such conduct he said that engineers were <u>well known to be a class of persons who looked upon money as dirt.</u>

The underlining is Brunel's. This, and the lack of punctuation, shows how strongly he felt about such insubordination.

Anxious at the rising costs of the enterprise, the tunnel directors tried to get some return on their investment by admitting visitors, at a shilling (5p) a time. Six or seven hundred sightseers descended the shaft each day and crowded into the tunnel. A concert was staged in the finished section; the acoustics were said to be particularly good. However, these were distractions that the miners and their supervisors could have done without. The ground was now so soft and silty that the men could scrape it away with their bare hands. Bones and broken china came through the riverbed overhead, and there was a constant influx of water. As Marc noted in his diary:

> *Notwithstanding every prudence on our part, a disaster may still occur.*
> ***May it not be when the arch is full of visitors!***

Isambard persuaded the directors to hire a diving bell from the West India Docks. This was an open-mouthed contraption, lowered on a chain, and fed with air from a pump on a support barge. He went down, poked around in the mud on top of the tunnel – and easily hit the top of the shield underneath. Some gold pins were passed through an iron pipe as a memento of this extraordinary moment. The final proof of the instability of the ground came several days later when a shovel that Isambard had left on the riverbed made its appearance inside the tunnel.

Thames watermen told stories of gravel dredging in that area of the river. Whatever the reason, the tunnel showed clear signs of imminent collapse. Isambard noted several false alarms in his journal. One was on 29 April, recorded under the marginal heading 'Terrible Panic':

> *Having got up rather late, Gravatt and I were at breakfast when Cook came running with a face like death to say that 'all was over', the Tunnel fallen in and only one man escaped. Gravatt being duped ran down immediately. I followed half-naked. Astonished to hear – nothing – ran on. Found the journeymen all gone but not seeing their bodies concluded they were safe – a long time discovering the cause of the alarm. At last perceived a slight run between 2 & 3, the silt having accumulated on floor plate. At last fell – splash!! Away – away. Watchmen to blame – all at breakfast but one.*

It did not take much for similar panics to start. Isambard and his assistants had to stay outwardly calm, but secretly they made themselves ready for the worst. Recalling the events of the night of 18 May, Richard Beamish wrote:

> *The visit of a dear friend (Lady Raffles) with a large party, about five o'clock p.m., did not tend to allay a strong feeling of apprehension which took possession of my mind. No sooner had she taken leave than I prepared*

▷ Engraving, from *Memoir of the life of Sir Marc Isambard Brunel*, by Richard Beamish, showing the diving bell in use under the Thames. Beamish noted, 'By dropping almost out of the bell we were enabled to place one foot upon the back of the top staves, and the other on the brickwork of the arch.'

myself for what, I was satisfied, would prove a trying night. My holiday coat was exchanged for a strong waterproof, the polished Wellingtons for greased mud boots, and the shining beaver for a large-brimmed south-wester.

His preparations proved justified at high tide that evening. First a trickle, then a torrent of water roared through the shield, sweeping the workmen towards the mouth of the tunnel. Isambard grabbed a rope and slid down one of the iron ties of the shaft to rescue an old engine man who was gasping for breath in the water. Having reached the relative safety of the staircase, they just had time to look back and see the dramatic spectacle of a stream of timber and other debris, eerily spot-lit by the gas lamps in the tunnel roof. A final crash marked the collapse of a temporary office, and the lights went out.

Repairing the damage took months. Isambard took another hazardous trip in the diving bell, where he found a hole that was later plugged with bags of clay interlaced with hazel rods. Soon, enough water had been pumped from inside the tunnel to enable him to examine it from a punt, crawling over a bank of silt that had built up over the shield. His private diaries show that he enjoyed the dangerous drama of these excursions, which attracted much publicity.

▽ 'First visit to the shield after the first irruption of the river'. Engraving from Beamish's biography of Marc Brunel.

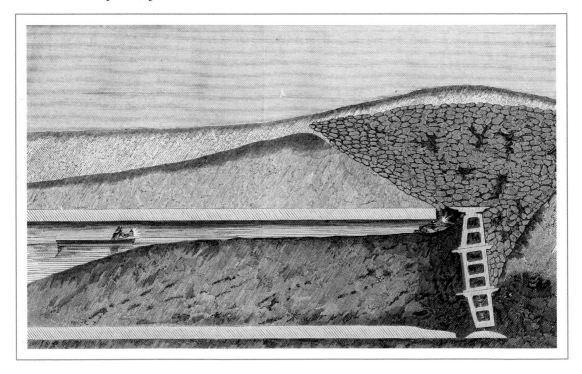

His *coup de théatre*, though, was the banquet held in the tunnel in November 1827 to celebrate a hard-won victory over disaster. To the accompaniment of music from the uniformed band of the Coldstream Guards (Beamish's old regiment), guests sat beneath arches hung with crimson drapery and lit with portable gas candelabra.

Two parallel celebrations were held. On one side of the arches were some fifty tunnel directors, engineers and members of the Establishment. On the other were more than one hundred workmen. At the High Table, toasts were proposed to the King, the Army, the Navy, and anyone else they could think of. For each toast, the band struck up an appropriate tune and the workmen joined in the echoing cheers. They, in their turn, honoured *their* heroes; towards the end of the evening, the foreman presented Isambard with a pick and shovel. One illustration – reproduced in L T C Rolt's biography – shows Marc Brunel being welcomed to the banquet by his son, but in fact he was not present. It was Isambard's show.

An artist had been invited to record the banquet, and a few days afterwards, Isambard went to view the results. In his journal, he complained that a waiter with a bottle of port had been placed in the foreground of one painting. *'Ye gods!! What refinement of feeling!'* he exclaimed. His reaction emphasizes that this event was as much about making engineering respectable as it was a celebration of man's mastery over nature. As Simon Schaffer explains:

'Dining in a tunnel, or staging a party on board a boat or in a railway carriage, showed that link between genteel society life and the wonders of engineering that Brunel really needed to forge. He needed to show that engineering was "polite" and powerful. A very good way of doing that was to have a candlelit dinner, tens of feet under the Thames, surrounded by mud and water, with all the waiters perfectly dressed and white linen tablecloths and aristocrats and gentry and patrons and entrepreneurs sitting around making polite conversation.

'There's an element of the joke there, as he's bringing together in one place two elements that aren't really supposed to meet – polite society and muddy engineering. But by bringing them together he shows his own mastery, that he's produced the social world by dominating the natural one.'

The Thames Tunnel publicity machine continued working as visits resumed. A vivid contemporary account, which shows Isambard in his role as impresario, appears in a letter to a friend from the actress Fanny Kemble:

The profound stillness of the place, which was first broken by my father's voice, to which the vaulted roof gave extraordinary and startling volume of tone, the indescribable feeling of subterranean vastness, the amazement and delight I experienced, quite overcame me.

Mr Brunel, who was superintending some of the works, came to my father [the actor Charles Kemble] and offered to conduct us to where the workmen were employed – an unusual favour, which of course delighted us all. So we left our broad, smooth path of light, and got into dark passages, where we stumbled among coils of ropes and heaps of pipes and piles of planks, and where ground springs were welling up and flowing about in every direction, all of which was very strange.

The appearance of the workmen themselves, all begrimed, with their brawny arms and legs bare, some standing in black water up to their knees, others laboriously shovelling the black earth in their cages (while they sturdily sang at their task), with the red, murky light of links and lanterns flashing and flickering about them, made up the most striking picture you can conceive.

As we returned I remained at the bottom of the stairs last of all, to look back at the beautiful road to Hades, wishing I might be left behind, and then we re-ascended, through wheels, pulleys, and engines, to the upper day.

By January 1828, working conditions were becoming increasingly hellish, and at high tide in the early hours of the 12th, the inevitable happened. Beamish had just come on duty, and was giving orders for rations of warm beer and gin to be given to the shivering workers going off shift. Isambard was in the frames, where he had been all night, when a column of soft ground suddenly gave away, and a torrent of water poured in. Historian Adrian Vaughan gives his version of what followed, based on his reading of Isambard's private diaries:

'Everybody runs except Brunel, because he is so fascinated by the sight of all this water thundering through the tunnelling frame with the candlelight glittering on the black water. Instead of running for his life, he stands there to enjoy it, which I think very much proves the dramatic side of Isambard's character. He didn't even think about his danger, not at first, and then he realizes, "Oh my goodness, the water's everywhere – I'd better get out of this."

'He turns to run for it and he is knocked down by a big piece of timber, which is being flung along by this torrent of the river coming in from above. He's knocked unconscious and he floats in along this tide of incoming water.

△ Engraving by an unnamed artist of water pouring through the shield in the fatal accident of January 1828.

'Meanwhile all the other men are pouring up the staircase out of the shaft, onto the ground above, and shouting, "The tunnel's broken in". They're jamming the door that leads out from the staircase. And Beamish, who is one of the site engineers, thinks, "I wonder if there are any people on the other staircase?"

'He hurries round, unlocks the other door, and just as he does so the water reaches its highest level, it comes up the shaft, and on top of it, face down, is a person. Beamish grabs this guy's jacket just as the water recedes. If he'd been two or three seconds later, the body would have gone. He pulls this guy out, lays him down, and it's Isambard.

'That is the first of two or three absolutely miraculous escapes. Fate had something in store for this man. If Beamish had been two or three seconds later he would definitely have gone and been drowned.'

According to Beamish, all Isambard could do for a while was to call out 'Ball! Collins!' – the names of the workmen who had been underground with him. But they were drowned, along with four others.

Once the bodies had been recovered, the next step was to use the diving bell once more to examine the damage to the riverbed. Despite severe injuries to his leg, Isambard insisted on being taken aboard the barge from which the bell was lowered, although he could do no more than whisper instructions from a mattress. He stayed at the scene for the rest of the next day, but was then overcome by illness.

Isambard's entries for this time in the Thames Tunnel Journal are as eloquent in what is left unsaid – the spaces between the entries – as the entries themselves:

Saturday 12 January
2 a.m. Went below little thinking how I should come up again – all going well.
Water
Taken Ill

Tuesday 15 January
Leg painful and numbed.

Wednesday 16 January
Felt worse. In the evening very unwell – dreadful headache. Dr Morris came down and took me up in his carriage – never felt so queer – obliged to be lifted in. Could not bear the least shake – felt as if I should be broken to pieces. Was cupped I believe, that night but don't remember.

Marc was left to work out how to block up the hole, with more than five hundred suggestions offered by members of the public. None of these proved any use, and in the end the water was kept out by the simple expedient of pouring in about 4,500 tons of clay and gravel. However, lack of money made it impractical to continue tunnelling. The shield was bricked in, a mirror was placed against the wall by the visitor's arch, and the tunnel became a curiosity for sightseers.

Isambard, meanwhile, turned out to have severe internal injuries that prevented a quick recovery. He went to Brighton to restore his health. It would be months before he was on his feet again – and many years before work on the tunnel could restart. By then, Isambard would have moved on to higher things.

▷ The Thames Tunnel was a magnet for sightseers, both during its construction and after its eventual completion. However, the horses and cart in this romanticized drawing are a product of the artist's imagination.

THE GREAT GORGE

While Marc and Isambard Brunel had been grubbing about in the mud underneath the Thames, an equally powerful engineering double act was being forged in the North of England. George and Robert Stephenson were about to establish themselves as the leaders of the new Railway Age.

George Stephenson was a rough, semi-literate Geordie, and a self-taught engineer. Any learning he had came from night school; his genius was in his hands, rather than his head, and he distrusted theory. It is hard to imagine anyone more different from Marc Brunel, the precise, educated Frenchman. However, they both shared a determination to give their sons – who were almost exact contemporaries – the best possible education.

Robert Stephenson – 'a wee sickly bairn', according to his biographer – was born in 1803. At the age of twelve, he was taken away from his village school and sent to the fee-paying Percy Street Academy in Newcastle. At first he walked there and back every day, a round trip of some 10 miles (16 km); after a while, though, his father bought him a donkey with which to make the journey.

At school, the young Robert's heavy boots, pit accent and rough clothes led to much teasing. However, school did gain him an education, which was matched with practical training when he was apprenticed to a mining engineer at the age of sixteen. This was hard and hazardous work, not ideal for a young man with weak lungs. By the age of eighteen, Robert's father had taken him out of the mines to become his assistant on the survey for the Stockton and Darlington Railway. This 25-mile (40-km) long railway, built to carry coal from the Durham coalfields to the mouth of the River Tees, became the first public line in the world to use locomotives.

Robert Stephenson's first biographer, J C Jeaffreson, paints a highly idealized picture of his new working conditions:

> *Spending the entire day in the clear balmy air, eating frugal meals of bread, butter, milk and potatoes under sheltering hedgerows, and lodging by night in roadside inns, George Stephenson and his assistants made holiday of their toil.*
>
> *When the survey was completed, and the map was plotted, Robert Stephenson's name was put upon it as 'the engineer', and no mention was made of his father. This was done at George's particular direction; and a more affecting instance of paternal devotion it would be hard to imagine.*

George undoubtedly gave Robert many advantages. He paid for his son to spend a term at Edinburgh University. He established the firm of Robert Stephenson and Co to build locomotives for the new line, thus kick-starting Robert's engine-building career when he was still only twenty. However, within six months of the company being established, Robert left England for South America.

This decision could simply have been due to the fact that, like many young men of the time, Robert was attracted by dreams of amassing wealth from the gold and silver mines there. It could also have been a reaction to his father's over-bearing personality. Whereas the young Isambard appeared to have a perfectly equable working relationship with his father – Marc's only concern being that his son was working too hard – Robert was constantly in his father's shadow.

Simon Schaffer highlights the contrast between the two father-son relationships:

'Marc Brunel works extremely hard in all sorts of ways to make a place for Isambard in the programme of engineering. The work in Paris, the work with Maudslay, is very much thanks to Isambard's father. Whereas the relationship between George and Robert Stephenson is stormy at the very

△ Previous page: Robert Stephenson as a young man. Engraved from a portrait by George Richmond.

▷ This romanticized Victorian painting helped to create the myth of George Stephenson, the self-made man. George is pictured in the centre, surrounded by his family. He is holding his miner's safety lamp; in the background is his first steam locomotive from Killingworth Colliery. Robert Stephenson is on the right.

best and Robert clearly works his way out from under the overwhelming presence of the great George Stephenson, the ultimate self-made man in early Victorian Britain.

'The relationship between Robert and George Stephenson is in many ways much more tense, much more neurotic, than the relationship between Isambard and Marc Brunel. And that's significant, because of the comparison between, I think, a rather conservative view of what engineering can achieve in the Stephenson programme and a rather visionary, not to say exaggerated view of what engineering can achieve in the Brunel programme.'

George was clearly a brilliant, intuitive mechanic. He had a talent for putting his finger on the heart of some engineering problem that had baffled those with more academic training. However, because his approach was practical, rather than theoretical, George tended to stick with what he knew, rather than working things out from first principles. The Stockton and Darlington Railway was the same gauge as a coal mine railway simply because that was the measurement George was used to and not because that was necessarily the best gauge for a railway track.

None of this would have mattered if it were not for the fact that, because George Stephenson built the country's first railways, this gauge became the standard for others to follow. The dispute that later erupted between the Brunel and Stephenson way of doing things, had its origins in George's innate conservatism.

Robert Stephenson's work with the Colombian Mining Association in South America covered roughly the same period as Isambard's struggles in the Thames Tunnel. Both shared the problems of any young engineer in asserting authority over those who could not accept that anyone who had not risen through the ranks knew more than they did. In Robert's case, he was not only honing his engineering skills – he was reinventing himself. He asked the few Englishmen he met to correct his heavy North Country dialect, so that he could pass for an English gentleman.

His return journey home, after a three-year stay, was marked by high drama when his ship was wrecked in a hurricane off New York. Robert and his companions landed safely, albeit with the loss of most of their money and luggage. The young Robert determined to become a Freemason after observing the way that a ship's mate had given priority to a fellow mason in the lifeboats.

Back in England, George was engaged in the spectacular engineering feat that was the Liverpool and Manchester Railway. Thanks to canal construction, people were reasonably familiar with the concept of cuttings, tunnels and embankments, but sceptics scorned George Stephenson's claim that he could cross the great Chat Moss bog. This swamp was so soggy that railway labourers had to tie planks to their feet to avoid being sucked in – and it was these planks that gave George the inspiration for his solution. He cut down a plantation of trees, and laid them together in a herringbone pattern to form a raft on which to float his rails.

The railway was almost built by the time the directors turned their attention to the form of propulsion to be used. With the wisdom of hindsight, it may seem astonishing that there should have been any doubt about this. However, many people were still unsure about the whole idea of steam locomotives, or 'travelling engines' as they were called. They preferred the tried-and-tested method involving stationary engines, which used ropes to haul waggons along the rails. This was the system found in many collieries, and on parts of the Stockton and Darlington Railway. In comparison, steam locomotives were considered to be clumsy and unreliable.

△ Robert Stephenson's *Rocket*, whose victory at the Rainhill Trials ensured a future for steam locomotives.

Eventually, it was decided to put the locomotive to the test in a publicly staged trial, with a prize of £500 for the winner. In theory, this was an open competition – but it could also be seen as a carefully manipulated exercise in public relations for Robert Stephenson and Co. Simon Schaffer compares the Rainhill Trials, staged over nine days in October 1829, to another popular spectator sport – prize-fighting:

'These quasi-human, quasi-animal objects blowing steam and smoke on open turf, waging war against each other, they are prize-fighters. They have reputations, they have heroes, and their pictures are on bar room walls throughout the country. That's as true of an engine as it is of a boxer.'

The trials, which attracted crowds variously estimated as 10,000 to 15,000 people, featured a bizarre variety of entries. One – the *Cyclopede* – was little more than a light cart powered by a treadmill worked by two horses. However, the front runners were clearly the steam engines – and the success of the *Rocket* firmly established Robert Stephenson's reputation as a locomotive engineer.

In a typical locomotive of its time, the firebox would generate hot gases and pass them through the boiler in a single tube to heat the water and produce the steam. However, Robert's engine, the *Rocket*, had a secret weapon

that lifted it above the competition. It used twenty-five separate tubes to carry hot gases through the boiler. This dramatically increased the heating surface area, which, in turn, increased the steam pressure. For the 60 miles (96 km) it had to cover during the trials, *Rocket* averaged a speed of 14 m.p.h. This may not seem particularly impressive now – but at the time, it was a landmark.

While Robert Stephenson was producing the first modern steam locomotive, Isambard was embarking on a project that would gain him vital contacts in his quest for fame.

The disastrous inundation in the Thames Tunnel had left him with internal injuries from which he took months to recover. He cannot have been an easy patient. A young man whose declared ambition was to 'Be the first engineer and example for all future ones', was hardly likely to enjoy languishing in bed.

His big chance came when he heard of a competition for a new bridge across the Avon Gorge in the wealthy Clifton area of Bristol. This was a project tailor-made for Brunel, for what could be more dramatic than throwing a bridge across a deep, yawning chasm?

◁ The old and the new.
A sailing barge on one of
England's earliest
canals passes
underneath the Sankey
Valley Viaduct, one of
many dramatic
engineering feats on
the Liverpool and
Manchester Railway.
Designed by William
Allcard, the viaduct was
erected under the
direction of George
Stephenson at a cost
of £45,000.

In 1753, an old Bristol wine merchant, William Vick, had left £1,000 in his will to be invested with the idea of building a bridge from the limestone heights of Clifton to Leigh Woods on the other side. His original plan was that, when the sum reached £10,000, the money would pay for a stone bridge, as well as compensation to the ferrymen who would be put out of business.

However, when the time came, the Ancient Society of Merchant Venturers, who had been entrusted with the money, found that £10,000 was nowhere near enough. Inspired by Thomas Telford's recently opened suspension bridge over the Menai Straits at Anglesey and by the idea that iron would be cheaper than stone, they decided to advertise for designs along similar lines.

By the closing date of the competition – 19 November 1829 – twenty-two plans had been submitted. Three were for stone bridges, ignoring the wording of the advertisement. One romantic entry from Birmingham featured rugged stone towers, styled to look like picturesque ruins. A slender iron bridge stood almost like an afterthought on top.

Isambard invested much time and effort into a total of four separate designs. He toured the area, painting and sketching. He spent two days in North Wales studying Telford's Menai Bridge. He also drew on Marc's practical experience as the designer of two iron suspension bridges for the little French island colony of Bourbon (now known as Réunion), 430 miles (700 km) east of Madagascar. It is true that these had a span of little more than 120 ft (36 m), whereas Isambard's most ambitious design for the bridge at Clifton was nine times that long. However, the Bourbon bridges had to be able to resist tropical hurricanes of up to 100 m.p.h. What they lacked in length they made up for in strength, and it is clear from Marc's diary that he had a considerable input into the calculations behind Isambard's designs.

The competition organizers had specified which section of the gorge they wanted the bridge to serve, but within these limits Isambard chose four different crossing points. His favourite design – called the Giant's Hole, from its nearness to the cave – sprang straight from rocks on either side of the gorge. Norman-style towers carried the suspension chains, and there were a number of precautions to stop the whole structure from swaying in the wind. This included using cross-bracing and inverted chains below the roadway – and chains brought down almost to the level of the platform at the centre of the span. No extra piers supported it from underneath; he felt this would add unnecessary expense.

Previous suspension bridges had used short connecting links between each loop of the chain; Isambard proposed replacing these with pins. Each link was to be nearly twice as long as the longest links on the Menai Bridge – and the bridge itself would be the longest suspension bridge built to date. This was truly a giant's bridge.

Isambard's artistically presented designs found favour with the competition jury, but they felt unqualified to make a judgement on their technical merits. Accordingly, they invited Thomas Telford – by then over seventy – to adjudicate. He dismissed Brunel's ideas as impractical, declaring that no suspension bridge should be longer than 600 ft (183 m). Anything over this length – and Brunel's favourite Giant's Hole design was 900 ft (330 m) long – was bound to be dangerous in a high wind.

Since 600 ft (or, to be exact, 579 ft) just happened to be the length of his Menai Bridge, and since Telford then went on to produce an alternative design of his own, it is tempting to dismiss his words as no more than those of an old man jealous of a young upstart. However, there is a history behind them.

Telford had started from humble beginnings. He never married, and his life was totally dedicated to his work. His only interest outside it – an unusual one for an engineer – was writing poetry.

He was born in the Scottish lowlands, in Eskdale. His father was a shepherd who died when Thomas was very young, so the boy was brought up by his mother alone. The family was poor, but had influential friends, and Thomas was able to get an apprenticeship as a stonemason. His first work was very practical, building little bridges; one survives, still with his mason's mark.

However, Telford decided his future lay in the city. From Edinburgh, he travelled down to London, where he worked as a stonemason on Somerset House. Then he moved to Plymouth, and eventually landed a job as County Surveyor in Shropshire.

From relatively modest projects, such as the renovation of Shrewsbury Castle, Telford progressed to designing iron bridges and entire canals. In 1801, he was commissioned by the government to survey Scotland's roads – and within two years, a massive programme of construction had started that was to revolutionize communications. In all, Telford was responsible for some 900 miles (1,450 km) of roads, together with the bridges that went with them.

Thomas Telford is probably best remembered today for his Conway and Menai bridges, and for the dramatic Pontcysyllte aqueduct. The name for this waterway, which simply means 'connecting bridge', is the ultimate in understatement. Anyone who has ever travelled on a canal boat across this narrow iron trough, which stalks across the Dee Valley on 120-ft (36-m) high

△ This formal portrait of Telford – commissioned by the Institution of Civil Engineers – underlines his Establishment status. It was painted by Samuel Lane, and shows the Pontcysyllte Aqueduct in the background.

stone piers, will know the dizzying feeling of looking over the sides to see what appears to be nothing between your boat and the ground below.

Telford's talents for finding the fastest, most economical and most artistically satisfying routes for canals, roads and bridges were recognized when he was made the first president of the Institution of Civil Engineers in 1820. As his biographer Anthony Burton puts it: 'In his last years he was *the* civil engineer in Britain.'

The Menai Bridge, which links the island of Anglesey with the Welsh mainland, marked a new era in the history of bridge building. Suspension bridges were not a new idea. Simple constructions of ropes and chains had long been used the world over. In 1817 a patent was taken out by the chain manufacturer Captain Samuel Brown for forming iron suspension bridges. However, the Menai Bridge was easily the most ambitious – and, at the time of its completion, in 1826 – the longest ever built.

Several suggested designs had been put forward for conventional bridges at this spot. All had been defeated by the Admiralty's insistence that even the tallest sailing ships must be able to pass through the straits unhindered. Telford's plan provided an elegant solution. Instead of using supporting piers on the riverbed, he built two towers on either side of the river. From the top of these were suspended the massive wrought-iron chains that held up the roadway.

Building work started early in 1820. It took up to three hundred men some four years just to build the piers and arches on either side of the straits. Telford had decided to build open arches, rather than closed rubble walls, to ensure that every cemented joint could be inspected as work progressed. All the stones in the piers, which were to carry the heavy weight of the suspension chains, were reinforced with iron dowels.

Several hundred experiments were carried out to determine the optimum measurements for the iron parts to be used in the bridge. Every stage of construction was carefully supervised – and, in a neighbouring valley, rehearsals were held for the great moment when the chains were to be raised. This involved improvising a chain of the same length, by fastening together fifty-seven of the 10-ft (3-m) long vertical suspending rods, and trying out various lifting methods. Eventually it was decided that the safest solution would be to load the central section of each chain on to a 450-ft (135-m) long raft, float it to the site of the bridge, and hoist it into place using capstans (chain-winding gear) and tackle.

On either side of the straits, men hacked out tunnels through the solid rock to anchor the chains. Each chain had a separate tunnel and all were connected to a solid iron frame deep into the rock. The chains were secured on top of the two towers and left to dangle down, ready to be connected. Then the first central section of chain, weighing 23.5 tons, was loaded on to a raft. All was ready for the momentous floating of the chains.

The Menai Bridge was part of an entire road system, the last link for mail coaches to cross to Holyhead in order to connect with the ferry to Ireland. It was work of national importance; it was also one enormous experiment.

As Anthony Burton puts it: 'It was a totally unknown technology – people had done it, but only on the smallest of scales. Can you imagine the huge amount of worry involved?'

Telford was prepared to take advice from scientists and mathematicians, but he was never really happy until he had tried things out for himself.

'He built test rigs,' says Anthony Burton. 'You simply test to destruction. You see how much pressure you can put on the damn thing until it breaks. When it breaks, you know that's as much as it can take. It's a very straightforward thing to do, but you still have to believe your own results.'

On 26 April 1825, the date set for lifting the first chain, Telford got to see whether his experiments would work on the larger scale. Samuel Smiles, in his biography of Telford, described the scene:

> *An immense assemblage collected to witness the sight; greater in number*
> *than any that had been collected in the same place since the men of*
> *Anglesea, in their war-paint, rushing down to the beach, had shrieked*
> *defiance across the Straits at their Roman invaders on the Caernarvon*
> *shore. Numerous boats arrayed in gay colours glided along the waters; the*
> *day being bright, calm, and in every way propitious.*

An hour before high water, the raft bearing the main chain was cast off from the shore and towed into place. One end of the chain was bolted to that hanging down from the pier on the mainland; the other was attached to ropes connected to two capstans fixed on the Anglesey side. The ropes were passed

△ Coloured aquatint showing Telford's road bridge over the Menai Straits, built to replace a treacherous ferry crossing. However, the winds that caused problems for boats also gave Telford some anxious moments with this early suspension bridge.

over blocks on top of the pier, with some 150 labourers on the capstans to haul them in:

> *When all was ready the signal was given to 'Go along'. A band of fifers struck up a lively tune; the capstans were instantly in motion, and the men stepped round in a steady trot. All went well. The ropes gradually coiled in. As the strain increased, the pace slackened a little; but 'Heave away, now she comes!' was sung out. Round went the men, and steadily and safely rose the ponderous chain.*

The fifers not only provided music while the labourers worked – they also had the very practical function of giving the men a beat to work to. The beat no doubt slowed to a dirge as the weight increased – but eventually the chain was raised up and the raft floated away. The same process was repeated until, finally, all sixteen chains were in place.

Samuel Smiles describes how Telford had sleepless nights during the final stages of the bridge's construction – and, after the successful raising of the chains, was discovered by friends down on his knees in grateful prayer. However, this was not the end of his troubles, as Anthony Burton explains:

'When it was first erected, the actual bridge platform was found to twist, and they realized it needed extra stays. It was said by the people who were working on it at the time that it was like being at sea. Some of them actually felt seasick because it was rocking and rolling – it was really quite a dangerous structure.

'It wasn't a difficult job to correct it. It's just that people hadn't realized that you needed to think about forces moving in different directions. He was thinking in terms of the weight of the bridge and wasn't thinking about lateral movement.'

Extra stays were put in to strengthen the bridge, and there was a triumphal opening in January 1826. Telford thought he had solved the wobble until, one windy wintry night, horses pulling a mail coach took fright. The coach overturned, and the horses were only prevented from falling into the straits below by their tangled harnesses. It seemed as if everything was in doubt once more. As Anthony Burton puts it:

'All the Jeremiahs came out and said, "Oh, we told you so, we said it would never work." Then the news came back – the bridge was fine. The driver, however, wasn't. He was blind drunk, and it wasn't the bridge – it was all down to the driver.'

One person who finds it very easy to sympathize with Telford's feelings is Chris Wise, the engineer in charge of London's most recent wobbly bridge. The Millennium footbridge, from St Paul's to Southwark, opened in a blaze of publicity – and then abruptly closed after developing an alarming tendency to sway under the weight of human traffic.

'On the Millennium Bridge, it was risky but we wanted to do a cutting-edge project, we wanted to do something that was absolutely of our time,' explains Chris Wise.

'We did everything we possibly could to check out every natural force that the bridge would be subjected to and it passed with flying colours. But because it was an experimental bridge, and because it was pushing the boundaries, we found ourselves building an object which would be exposed to natural forces the like of which a bridge had never seen before. As a result of that it responded in a way which had never been seen before, and caught us out.

'Telford's big bridge did exactly the same. If you imagine it like a giant guitar string, when you blow sideways on it with the wind, it wobbles. Because it was a very long guitar string – a very long bridge – it wobbles very slowly. There are all sorts of little funny eddies and things in the wind which can actually get to be quite rhythmical. And that's not very good.

'He wouldn't have known that in advance. They didn't have wind tunnels. The only way they could ever find out these things was to build something and see what happened to it. In that regard it's exactly the same as the guys who built the Gothic cathedrals.

'What was really good about it was that he had the courage to get it built in the first place – and when it went wrong, he fixed it.'

The Menai Bridge has stood for over 170 years, bearing a weight of traffic unimaginable in Telford's time. However, at the time he was asked to judge the Clifton Suspension Bridge competition, he had good reason to be cautious. As Anthony Burton puts it:

'I think he genuinely had a worry that the scheme that was put before him was not safe, that it was a dangerous plan and that people would lose their lives if it was built. He really thought that the bridge that Brunel designed would not stand up, it would collapse.'

Having declared none of the designs entered for the Clifton Bridge competition to be suitable, Telford was invited to produce his own. The monumental folly that came from his drawing board could not have been more different from the sleek, elegant lines of his Menai Bridge. L T C Rolt describes it as 'the one truly monstrous aberration of his long career', his only possible excuse being senile decay. However, Anthony Burton puts forward a plea in mitigation:

'He had this secret thing about being an architect. When he was a surveyor back in Shropshire he actually designed churches and he was bitten by the Gothic bug that was then around. He'd built a Gothic aqueduct in Birmingham, and he thought, "Wow, now here's something I can do that's on a really big scale".'

Big it undoubtedly was; a three-span suspension bridge supported upon two elaborate gothic piers rising through the full height of the gorge. Each pier was hollow, with a gallery halfway up reached by a flight of stairs. The central span was 360 ft (110 m), with two side spans of 180 ft (55 m) each. It looked as though Telford's obsession with lateral wind resistance had led him to abandon all sense of fitness for purpose.

Brunel wrote sarcastically to the Bridge Committee:

*As the distance between the opposite rocks was considerably less than what
had always been considered as within the limits to which Suspension Bridges
might be carried, the idea of going to the bottom of such a valley for the
purpose of raising at a great expense two intermediate supporters hardly
occurred to me.*

Telford's design was included in the prospectus issued by the bridge
directors when they put the bill before Parliament. However – in part, at
least, due to the expense – enthusiasm cooled. The directors decided on a
second competition, for which Brunel submitted his Giant's Hole design,
altered to include a stone abutment rising from the rocks on the Leigh
Woods side of the gorge. This reduced the central span to the required
600 ft (183 m).

A new panel of judges was appointed, and Brunel's design placed second.
The chief judge, Davies Gilbert (a past president of the Royal Society)
expressed reservations about some of the details. However, miraculously,
he changed his mind overnight. This was not to be the last time that Brunel
was able to use his charm and considerable powers of persuasion to swing
a vote in his favour.

Gilbert was an MP who had contributed several important modifications
to the bridge plans when they came up for parliamentary discussion. His
objections were technical ones, about the use of single pins for chain links,
the method of attaching the suspension rods, and of anchoring the chains to
the rocks. The moment Brunel heard of the committee's decision, and the
reasons behind it, he arranged a meeting with Gilbert. Brunel explained his
calculations in detail – and succeeded in getting Gilbert to announce his
design as the winner. There is no record of what the original winner – W
Hawks – thought of this coup. However, the cock-a-hoop tones of Brunel's
writings at this time tell their own story.

The design as eventually accepted featured massive gateways in Egyptian
style. The towers, guarded by sphinxes, were to be cased in cast iron,
ornamented from top to bottom with panels representing the construction of
the bridge. It all sounds disturbingly like a Stalinist Soviet monument to
the workers. At this distance, though, it is hard to judge. The original
sketches by the artist John Horsley have not survived, and the bridge as
we know it is considerably plainer.

△ Telford's cathedral-
like design may have
satisfied his unfulfilled
architectural ambitions,
but it was never built.

Brunel's private diary for 26 March 1831 recorded simply: 'Attended Committee. Unanimous in favour of Egyptian'. The following day he wrote at rather more length to his brother-in-law, Benjamin Hawes:

I have to say that of all the wonderful feats I have performed since I have been in this part of the world, I think yesterday I performed the most wonderful. I produced unanimity amongst fifteen men who were all quarrelling about the most ticklish subject – taste.

The Egyptian thing I brought down was quite extravagantly admired by all and unanimously adopted; and I am directed to make such drawings,

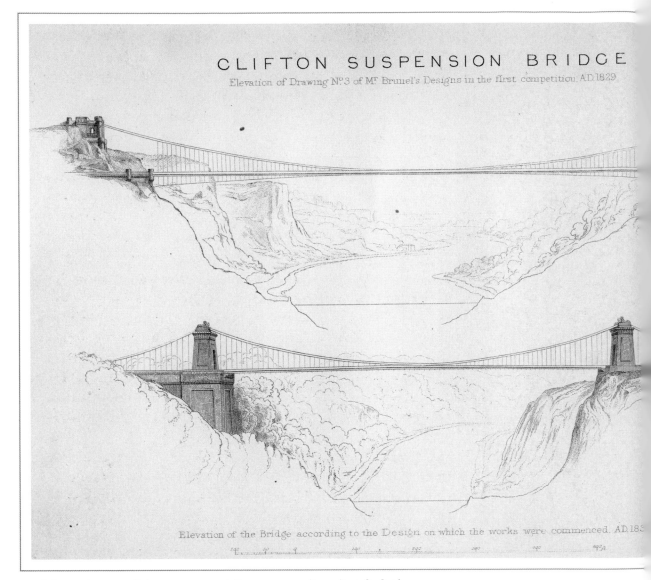

CLIFTON SUSPENSION BRIDGE

Elevation of Drawing Nº 3 of Mʳ Brunel's Designs in the first competition. A.D. 1829

Elevation of the Bridge according to the Design on which the works were commenced. A.D. 183

lithographs, etc as I, in my supreme judgement, may deem fit; indeed, they were not only very liberal with their money, but inclined to save themselves much trouble by placing very complete reliance on me.

His jubilation was to prove rather premature. Despite an elaborate stone-laying ceremony, the directors found it increasingly hard to raise the £57,000 needed to build the bridge. In the autumn of 1831, work was interrupted by the notorious Bristol Riots. A mob of citizens, furious about the slow progress of parliamentary reform, plundered public buildings, leaving a trail of fire and destruction. Brunel was sworn in as a special constable, and took an active part in rescuing some of the treasures from the ransacked Custom House. In such a climate, building a bridge for ladies and gentlemen of leisure to cross from one side of a gorge to the other was hardly likely to find favour. The project quietly died; the bridge was not to be built in Brunel's lifetime.

Within months, all of Brunel's plans seemed to have turned to ashes. As he wrote in his diary: 'So many irons, and none of them hot.' He produced plans for modernizing the Bristol Docks, but – like the Clifton Bridge – these were held up by the riots. He had designed an observatory with mechanical shutters for a client in Kensington. However, he had ended up being embroiled – not for the last time in his career – in wrangles about the cost. The Thames Tunnel was still closed. On 17 December 1831, under a diary entry dramatically outlined in black, he wrote:

> *Tunnel is now I think DEAD…The commission have refused on grounds of security – This is the first time I have felt able to cry at least for these ten years. It will never be finished now in my father's lifetime I fear. However nil desperandum has always been my motto – we may succeed yet – perservation.*

That same month he had taken his first ride on the Stephensons' Liverpool and Manchester Railway. Interleaved into his diary is a small sheet of paper bearing a series of wavering circles and lines with the note: 'Drawn on the L and M railway 5.12.31'. Next to it is written:

> *I record this specimen of the shaking on Manchester railway. The time is not far off when we shall be able to take our coffee and write while going noiselessly and smoothly at 45 miles per hour – let me try.*

He would soon get his chance.

△ Above – Brunel's favourite 'Giant's Hole' design for the first Clifton Suspension Bridge competition. Below – his final version, with Egyptian-style towers topped by sphinxes.

THE GREAT WESTERN: 1

The first sign of the massive project that was to dominate the next stage of Brunel's working life were two letters next to a diary entry: BR.

It was not long before the modestly named Bristol Railway became, in Brunel's hands, something rather more grandiose. By August 1833, BR had become GWR – the Great Western Railway.

'"Great" was a Brunel family word,' explains railway historian Adrian Vaughan. 'When Marc Brunel made a 10-ft [3-m] diameter circular saw, which was some engineering feat, he didn't call it his "grand saw", he called it his "great saw". Whenever he did something particularly large, it was always "great".

'If you look at any other engineering work of that period, it would be called "grand". Grand Junction Railway, Grand Union Canal. Nobody used the word "great". It's very Brunel. This is the actor in him, this is the artist.

'Brunel had always wanted to build the greatest railway that ever there had been. He constructed this remarkably relatively level railway, so much so that it earned for itself the nickname of "Brunel's Billiard Table". And to this day they're running trains at 125 miles an hour over it, without having to alter it in any way.'

However, Brunel almost lost the chance to design his dream railway when he took one enormous gamble. At the beginning of 1833, a committee had been formed in Bristol with the object of commissioning a preliminary survey of a railway line to London. The Bristol Docks were losing trade to Liverpool and the city's merchants hoped that a new railway line – like that from Liverpool to Manchester – would help to redress the balance.

Brunel heard that the committee was planning to organize a competition to find the cheapest route – but he wrote to them, saying he thought they were wrong:

> *You are holding out a premium to the man who will make you the most flattering promises, and it is quite obvious that he who has the least reputation at stake, or the most to gain by temporary success, and least to lose by the consequence of a disappointment, must be the winner in such a race.*

He threatened to withdraw his name. He would survey only one road; it would not be the cheapest, but it would be the best.

His gamble paid off, and – by just one vote – the committee agreed to appoint him Chief Engineer. At the age of only twenty-six, he was put in charge of the most ambitious railway the country had ever seen. He had just three months to make a preliminary survey of the entire stretch from Bristol to London before going public.

Although he had two assistants, this was an arduous time for Brunel. It was not simply a matter of taking a ride from Bristol to London. In a pre-aeroplane era, with no way of undertaking an aerial survey, all the work had to be carried out on the ground. He needed to climb every available grassy knoll and hill to gain a bird's-eye view. He also had to contend with landowners, many of whom did not take too kindly to having trespassers on their land.

Up at five or six most mornings, he surveyed by day, writing up his reports, estimates and calculations by night. He travelled on horseback or by coach, staying in inns or hotels. When he had to go longer distances he travelled at night. His diary entries for this period are essays in frustration – riding accidents, delays, crowded or full coaches and missing assistants:

> *Monday 22 April – Started at 6 a.m. Examined the ground in the neighbourhood of Wantage – breakfasted at Streatley. Determined on the outer line winding round the undulating ground. Returned to Reading, dined, and went to Theal to meet Hughes* [an assistant]. *After waiting some time gave it up and returned.*

△ Previous page: A young and confident Brunel with a plan of the railway that was to make his name. Oil painting by John Horsley.

Tuesday 23 April – After breakfast went in search of Hughes; after some trouble found him at the canal between Shinfield and Calcott Mills, a beautiful place this in hot weather. Gave him the line to Wantage.

The comment about the beauty of Hughes's chosen spot jars; it does not sound as if Brunel allowed himself much leisure time to appreciate such things. Just as he had spent nights and days in the Thames Tunnel, so he tackled this new project with the monomania of a driven man. This was his great chance to prove himself.

'The Great Western Railway route is an invisible masterpiece, because it's so big people don't notice it,' says Adrian Vaughan. 'But the way that he designed the route is definitely his first great stroke of genius.

'There were some mediocre engineers in Bristol who were trying to build this railway, who luckily for us never got the job. If they'd had their way, the railway would have followed the turnpike road. It would have gone up the Avon Valley to Bath and then followed all the river valleys – Bradford-on-Avon, Trowbridge, Devizes, Marlborough, Hungerford, Reading. That's the parochial way.

'Brunel had a wonderful eye for the lie of the land, and was a great strategist. He would probably have done very well in the Army. By going the way he went, he left his way completely open to throw off his branch lines or secondary main lines, to South Wales and into the North of England.

'If he'd gone the mediocre way, the shortest way, he would have ended up 650 ft [200 m] above sea level in the middle of the Marlborough Downs, and he couldn't have thrown anything off anywhere. If you try to go to the north or west from Marlborough you have to negotiate the escarpment of the chalk Downs, you're going to be on 1-in-40 gradients. So Brunel's great genius was to design the route the way he did it – it really was wonderful.'

Surveying the line was only the first stage. Brunel spent a considerable amount of time chatting up potential objectors. Country squires saw railways as an iron invasion of their feudal strongholds; city landlords saw nothing but a disastrous drop in property values. If Parliament gave permission for the railway, land could be bought by compulsory purchase and the owners compensated. However, it would smooth the bill's passage through Parliament if landowners could be placated in advance.

Money had to be raised by selling shares to investors. After the first public meeting, Brunel wrote:

Got through it very tolerably, which I consider great things. I hate public meetings: it is playing with a tiger, and all you can hope is, that you may not get scratched, or worse.

The title 'Great Western Railway' was formally adopted in August 1833, and a company formed with committees in both London and Bristol.

Nothing could be built without the consent of Parliament – yet, despite the fact that a railway linking the two major ports would be a huge national asset, all the money had to be raised privately. Parliamentary Standing Orders laid down that at least half the capital had to be raised before any bill could be put forward.

Brunel's estimate for the cost of the line was slightly over £2,800,000 – a fantastic sum in those days. Not surprisingly, by the end of October 1834, the directors had only managed to scrape together a quarter of the sum needed. However, they could not afford to wait until the next parliamentary session came round – so they decided to go ahead with a bill for only a part of the railway. Their plan was to make a start on the London and Bristol ends, and come back later for permission to link up these two sections.

Following the opening of the Liverpool and Manchester Railway in 1830, railway speculation was growing throughout the country. The London and Birmingham Railway, with Robert Stephenson as Chief Engineer, was by now well underway – and other companies were in competition for sections of the line to Bristol. Quite apart from the spirit of friendly rivalry that was to develop between Brunel and Stephenson, there must have been a strong sense that there was no time to lose.

The GWR bill passed through its first reading in the House of Commons relatively painlessly, although the idea of a branch to Windsor had to be dropped because of violent opposition from Eton College. Then the bill went to the committee stage, where proceedings were to last no less than fifty-seven days. A variety of engineering experts were called, including George Stephenson, who testified that he did not know of any existing line so good as that proposed by Brunel. Although these men were potentially in competition, when it came to defending a fellow engineer against the ignorant public, they stuck together.

Brunel, whose evidence lasted eleven days, was the chief engineering witness. Watching the skill with which he parried objections became a popular spectator sport. Robert Stephenson, who had undergone a similar grilling during the passage of the London and Birmingham Railway bill, was in the audience on at least one occasion. The earliest letter to survive in a lifetime's correspondence between the two men is a formal, hand-written note from Stephenson to Brunel, complimenting him on his performance.

Of course, only written records of Brunel's evidence survive, but Simon Schaffer suggests that it was delivered with more than a touch of Gallic charm:

'If you look at the journalists' reports about Brunel's speeches, it's obvious that he was a pretty impressive and compelling speaker,' he says.

'I think there's a sign of a very strong and rather seductive accent there – he's in some ways the Maurice Chevalier of British engineering. But on the other hand, he's capable of understanding the interests of different sections of the audience. This is a man who, if he hadn't been an engineer, would have been an extremely effective politician.

'It's an age when politics is about face-to-face interaction. There's no broadcasting – one meets face to face the people who count. And Brunel was obviously, close up, an extremely impressive figure who dresses very self-consciously, who takes great care with his public image, and who is a kind of seductive presence on the financial and economic stage of that time.'

Securing the passage of the bill through Parliament meant schmoozing with friends and opposition alike, and Brunel took rented rooms at Parliament Street to be close to the action. The GWR counsel, Sir George Burke, QC, lived opposite, and Brunel set up a communication system using a bell and string across the street. He would use this relentlessly at all hours of the day and night, and would often work into the small hours, falling asleep in his chair with a cigar in his mouth.

Much opposition to the railway came from vested interests – bridge-owners who would lose revenue from tolls, canal and stagecoach companies that would lose trade. Farmers near London opposed it, fearing competition; on the other hand, country farmers were delighted with the prospect of the new markets that would open up with a more efficient means of transport. Canals were notoriously unreliable. Quite apart from the slowness of horse-drawn barges, frosts, winter floods and summer droughts caused great delays.

Patient under cross-examination, Brunel established a clear need for the new railway. However, this first attempt at getting the bill through Parliament was to founder at the last moment.

The line as set forth was to terminate on the north side of the River Thames at Vauxhall Bridge. The last section was to have been carried on a 24-ft

(7.3-m) high viaduct, with a parapet 6 ft 6 in (2 m) high to prevent passengers peering into the windows of the elegant mansions around Sloane Square, Belgravia and Pimlico. Since the owners of many of these elegant mansions – some only very recently built – were influential members of the House of Lords, the GWR directors got cold feet halfway through the hearing. They decided to abandon the last 2 miles (3 km) of the viaduct. Instead, they made provision for a London terminus behind a pub called the Hoop and Toy, near the site where South Kensington tube station now stands. Needless to say, the residents of fashionable Brompton were not best pleased.

In the end, the bill was thrown out in the House of Lords. The main objection was that – with only the London-to-Reading, and Bath-to-Bristol sections built – it would be only the head and tail of a railway line. There were sarcastic comments that it was 'neither great, nor western, nor a railway'.

Undaunted, the directors went back to Parliament in March 1835. They had fresh proposals, having managed in the interim to raise the money necessary to put in a bid for the whole line. Bearing in mind the sensibilities of the genteel residents of Brompton, they suggested a terminus shared with Robert Stephenson's London and Birmingham Railway at Euston.

Brunel was saved from having to go through the whole eleven days of his previous evidence, although he still had some crackpot objections to contend with. The hottest debate centred on his plans for a $1\frac{3}{4}$-mile (2.8-km) tunnel, with a 1-in-100 gradient, through a hill at Box, near Chippenham. Chief among his opponents was the improbably named Dr Dionysius Lardner. Simon Schaffer explains:

'Dionysius Lardner was one of the chief railway journalists of the period. Like a lot of pundits, he had an exaggerated sense of his own expertise. He kept on confusing the fact that he could take a good brief with the fact that he knew what he was talking about, which he very often didn't.

'He published widely about the new inventions of the 1830s and 1840s. He got a university job on the strength of this, as one of the first professors of the public understanding of science in Britain. But his technical expertise was patchy.

'He was deeply concerned – or he pretended to be deeply concerned – about the slope of the rail through the long Box tunnel. He claimed that if the brakes on the train gave way when it was going through the tunnel, it would emerge at the other end doing more than 120 miles per hour.

'Brunel and his allies simply had to point out to Lardner that he'd forgotten about air resistance and friction completely. He'd committed a mistake that a schoolboy would have been ashamed of making.

'Lardner wasn't alone in making silly remarks about the feasibility of Brunel's schemes. William Buckland, who was one of the leading geologists in Britain at the time, a professor at Oxford University, said that Brunel's tunnels would fail. Every time a train went through them, the shaking would disturb the fragile strata and the tunnels would collapse.

'So there was the usual degree of incompetence among Britain's intellectual elite when faced with new challenges being posed by engineering systems. On the other hand, even though it's easy to make fun of men like Lardner and Buckland, it's also important to remember how crazy a lot of what Brunel was doing must have seemed at the time. We mustn't let the fact that Dionysius Lardner was obviously weak on the basic principles of physics blind us to the fact that Brunel himself would have seemed visionary and crazy to most of his audience.'

The Act for the Great Western Railway finally gained the Royal Assent on 31 August 1835. A fortnight later, almost as an afterthought, Brunel informed the directors that he wanted to construct the line to a new gauge almost twice the width of most other railways in Britain. He had conveniently manipulated events so that the standard width of 4 ft 8½ in (1.4 m) was not stipulated in the Act. This was Brunel's chance to make his railway bigger and better than any that had gone before. As Simon Schaffer explains:

'One of the most important features about Brunel's attitude to the whole Great Western Railway project is that he sees it as a coherent system. He doesn't see it as a series of separate puzzles, each of which can be or should be solved separately.

'That vision is there from the beginning, right through to its completion. It's not just that Brunel is going to be committed to wide gauge, to 7-ft [2.1m] spacing for the rails – that's part and parcel of a whole systematic view of what the railway is going to be.

'You see this very well in most of the famous anecdotes about the way the railway is built. So, when the more timid Bristol investors ask whether it's a good idea to build the railway from Bristol to London all in one go, Brunel's answer is, "Yes, and we should be able to engineer a transport system to carry people from London via Bristol to New York." The vision is global.'

The first railways were powered not by machines, but by horses. The rails were just wide enough for a single horse to walk comfortably between them and pull a coal waggon out of a mine. George Stephenson had designed the

Stockton and Darlington Railway to the gauge that happened to be in use at Killingworth Colliery, where he worked. With the addition of just half an inch (1.25 cm), he used the same measurement on the Liverpool and Manchester Railway. Most other railway pioneers had unthinkingly followed suit, and it had become the national standard.

However, to Brunel a gauge based on the hindquarters of a horse was unscientific. His engines would run on a much wider track. This would allow the carriages to be set low between the wheels, giving them a low centre of gravity and a more stable ride. His trains would have more space for passengers, and have bigger and faster engines.

He set out his reasoning in a letter to the GWR directors, dazzling them with his detailed technical calculations. At the end of the letter, he listed all the possible objections to his 'broad gauge'. The last of these was the inconvenience of making a junction with Robert Stephenson's standard-gauge railway at the London end. He concluded:

> *This I consider to be the only real obstacle to the adoption of the plan; one additional rail to each railway must be laid down. I do not foresee any great difficulty in doing this, but undoubtedly the London and Birmingham Railway Company may object to it, and in that case I see no remedy, the plan must be abandoned. It is therefore important that this point should be speedily determined.*
> *I am, Gentlemen,*
> *Your obedient servant,*
> *I K Brunel*

Brunel got his way, and the directors authorized the adoption of the broad gauge. In the event, negotiations with the London and Birmingham Railway foundered over the conditions on which GWR was to use the land at Euston, and a station was built at Paddington instead. Brunel got together a team of assistants, and work started at both the Bristol and London ends of the railway.

Not surprisingly, given the pressure of events, Brunel's diary-keeping had lapsed during this period. However, on Boxing Day 1835 he took up his pen once more:

> *What a blank in my journal, and during the most eventful year of my life! When last I wrote in this book, I was just emerging from obscurity. I had been toiling most unprofitably, at least for the moment, at numerous things.*

The Railway was certainly being thought of, but still very uncertain. What a change! The Railway is now in progress. I am now engineer to the finest work in England. A handsome salary of £2,000 a year, on excellent terms with my directors, and all going smoothly. But what a fight we have had, and how near defeat, and what a ruinous defeat it would have been! It is like looking back upon a fearful pass, but we have succeeded.

Brunel's 'handsome' salary, incidentally, set the standard for his contemporaries. As soon as the directors of the London and Birmingham Railway heard what he was being offered, they gave Robert Stephenson a £500 pay rise to match.

Brunel now had all the work he could handle. He had wowed the parliamentarians with his bravura performance; he had the faith of the GWR directors (for the moment, at least). He had the money to move to more elaborate offices and to set himself up in London society. Now all he needed was a trophy wife.

At the age of twenty-one, Brunel had confided to his secret diary, which he always kept locked, that he expected to remain a bachelor. As he put it: 'My profession is, after all, my only fit wife.' At one time, he had been in love with a Manchester girl called Ellen Hulme. However, she seems to have faded out of the picture in favour of the daughter of a musical Kensington family, Mary Horsley. Several commentators have described Mary Horsley as little more than a vacuous clotheshorse. She had classic beauty, but did not appear to have inherited the family musical talent. (Her father was an organist and music teacher, and Mendelssohn and other famous composers of the day were regular visitors to the household.)

Brunel's parents had married for love; theirs was a romantic story, set against the background of the French Revolution. The young Marc had been forced to flee France, leaving his English sweetheart, Sophia Kingdom, imprisoned in a commandeered convent. It was only after they were reunited in England that they were able to marry.

Some of Brunel's contemporaries – Robert Stephenson among them – married for love. However, Simon Schaffer suggests that, in marrying Mary Horsley, Brunel was simply following the example of others within his social circle. Men like Charles Babbage, the designer of an early form of computer, or Humphrey Davy, the chemist and inventor of the Davy Safety Lamp. As he explains:

◁ This portrait, like the
one with which this
chapter opened, was
painted by Brunel's
brother-in law, John
Horsley. The lines etched
on Brunel's face show
the effect of his
workaholic lifestyle.

'It was common for men of Brunel's class in the middle of the 19th century
to use marriage as another tool for social advancement. As these men saw it,
the function of their wife was to be their social secretary and administrator.

'Brunel wanted engineers to be recognized as gentlemen. One of the most
important places that a man like Brunel could recruit the cash he needed
for his projects was over the dining table, or in the salon, or at a party. That
meant having the wealth, the status, and the social manners to run affairs like

that. For him, everything was going to be sacrificed to this kind of social and technical ambition.'

Whatever the truth of their relationship, Brunel was a fairly absent husband from the moment they moved into their new home in Duke Street, Westminster. (A location chosen for its proximity to the House of Commons.) As L T C Rolt puts it, Mary's life was rather like that of the wife of a sea captain. While Brunel spent long hours travelling, she stayed at home, parading in St James's Park and supervising the household. She appeared with him on ceremonial occasions, but for the most part led her own life.

'What little we know of Brunel's private life suggests that he was a very engaged father when he was there, which wasn't very often,' says historian Andrew Lambert.

'There are wonderful set-piece cameos of *pater familias* Brunel, but they're mostly about Brunel. They're not about the children or about the wife. It's Brunel putting on a public performance for his children to write down and record in their memoirs later.

'In some senses he's always on display. You don't feel that he's relaxing at home. He doesn't shut the front door and take off his stovepipe hat and sit down and think, "Well, I don't have to do this any more."

'I think he was a public performer from the beginning of his conscious life. He was put on display to a certain extent by his father; he was in the company of learned men as a small child, and he never outgrew that. So he is always engaged, he never relaxes, he doesn't sleep very much – they fit together.'

Delegating was never Brunel's strong point, and from the start of operations on the Great Western Railway, he wanted to be everywhere at once. To avoid the exhaustion of long stagecoach journeys, he had his own *britska* – a large horse-drawn carriage (the word is Polish) – specially designed for his needs. It had a removable folding hood and enough space that he could lie down on long journeys. There was a bed, a drawing board and cupboards for surveying materials – plus a box that held 50 cigars. The navvies who worked on the line – and presumably dreaded its appearance – nicknamed it 'The Flying Hearse'.

By contrast, Robert Stephenson maintained a clear chain of command on his London and Birmingham Railway. He parcelled up the line into five sections; each was put in charge of a Resident Engineer, with three sub-assistants under him. Construction started at the London and

Birmingham ends, and there were never fewer than 12,000 men working at any one time. Stephenson moved from Newcastle to London so that he could supervise the project more easily – but his assistants had much more autonomy than those employed by Brunel.

The engineer F R Conder, who was Robert Stephenson's pupil, summed up the difference in the two men's working methods in his memoir, *Personal Recollections of English Engineers*:

> *The staff of Mr Stephenson, although it cannot be said to have had a military organisation, yet to some extent resembled an army corps in division and subordination of duty. Each officer was the servant of the Company. Each had his own limit of function and responsibility; and although Mr Stephenson well knew how to show, from time to time, that not the smallest details would escape his attention, if it involved what was wrong, the order of his office was not such as to overburden the Engineer-in-Chief with details that fell properly within the competence of the Residents, or even of the subs. But the Engineers on the broad-gauge lines appeared to regard themselves less as the officers of the Company than as the channel of the will of Mr Brunel. His Residents no more ventured to act without his direct authorisation, ad hoc, than did any inspector on the line.*

Like Stephenson, Brunel employed Resident Engineers. But despite the fact that these were talented men who went on to forge independent careers for themselves, Brunel did not seem to expect them to show much initiative. This applied to the designing of the railway in particular. Even the smallest details were sketched out in Brunel's hand. As his office expanded, he employed draughtsmen to transform these sketches into detailed drawings.

The grand design behind Brunel's and Stephenson's railways alike was based on technology developed during canal construction. The Birmingham and Liverpool Junction Canal had been designed by Thomas Telford towards the end of his career, as a challenge to the new railways. It proved to be a last fling for a dying form of transport. However, the railway pioneers copied its deep cuttings and embankments, and the contractors who built them simply transferred their skills.

We are now only too familiar with the heaving mounds of earth thrown up by motorway building projects, so it is hard for us to recapture the sense of wonder the ordinary person would have felt at the sheer scale of the operations involved in building the railways. Perhaps the best-known description comes from *Dombey and Son*, by Charles Dickens. Here he describes the scene of devastation caused by the London and Birmingham Railway as it carved its way through Camden Town:

> *The first shock of a great earthquake had, just at that period, rent the whole neighbourhood to its centre. Traces of its course were visible on every side. Houses were knocked down; streets broken through and stopped; deep pits and trenches dug in the ground; enormous heaps of earth and clay thrown up; buildings that were undermined and shaking, propped by great beams of wood. Here, a chaos of carts, overthrown and jumbled together, lay topsy-turvy at the bottom of a steep, unnatural hill; there, confused treasures of iron soaked and rusted in something that had accidentally become a pond. Everywhere were bridges that led nowhere; thoroughfares that were wholly impassable; Babel towers of chimneys, wanting half their height; temporary wooden houses and enclosures, in the most unlikely situations; carcasses of ragged tenements, and fragments of unfinished walls and arches, and piles of scaffolding, and wildernesses of bricks, and giant forms of cranes, and tripods straddling above nothing.*

The description continues with tales of 'hot springs and fiery eruptions, the usual attendants upon earthquakes'; it concludes on a note of sarcasm:

> *In short, the yet unfinished and unopened Railroad was in progress; and, from the very core of all this dire disorder, trailed smoothly away, upon its mighty course of civilisation and improvement.*

Once people realized how much money was to be made from the new railways, hostility began to diminish. However, there are tales of farmers ploughing up the surveyors' wooden pegs 'by mistake', and of landowners making extortionate demands for compensation. Every mile of every new railway was an exercise in compromise. The trick was to keep landowners happy without having to make massive, unnecessary loops simply to take the railway away from their front gates.

The letter books that contain Brunel's GWR correspondence show the level of detailed negotiations involved. He agrees to compensate a country rector for the loss of a fish pond; to tunnel underneath someone else's orchard; and to move the line 50 yards (46 m) further away from a fence belonging to a Miss Payne. However, some of the wealthier landowners he

▽ Drawing by John Bourne of the London and Birmingham Railway as it burrowed its way under Park Street, Camden, in 1836. This was the 'earthquake' described by Charles Dickens in *Dombey and Son*.

dealt with made more extravagant demands. Passengers on the Cheltenham branch of the GWR have to travel through the 415-yard (380-m) long Kemble Tunnel, because of Mr Robert Gordon's insistence that he should not be able to see the railway from his window.

When work started at the London end of the GWR, the terminus was temporarily established west of what is now Bishop's Bridge Road, behind the present Paddington Station. Unbelievable as it may seem today, the area now dominated by the elevated Westway motorway was then open country. There were cottages scattered in fields, and one or two more substantial residences. One of these, Westbourne Place, was occupied by Lord Hill, Commander in Chief of the British Army. His name survives in Lord Hill's Bridge, by what is now Royal Oak station.

Building materials were transported by canal, and work started on some of the great engineering feats that were to distinguish the railway. This included the 65-ft (16.5-m) high Wharncliffe Viaduct at Hanwell, named after the lord who had helped ease the GWR bill's passage through the House of Lords. It was decorated with details in the fashionable Egyptian style, as had been planned for the Clifton Suspension Bridge.

For Brunel, God was in the detail. The National Archive at Kew still possesses his Books of Facts, in which he noted down anything that might come in useful. This could be something as apparently insignificant as the annual rainfall for Bristol, or the type of grass species best suited to railway embankments.

This passion for detail can also be tracked in the GWR letter books. On New Year's Day 1838, Brunel wrote to one Edwin Maw from his home in Duke Street.

Dear Sir,

I am very anxious to hear from you to know what you are doing in the way of wood screws.

I have lately been compelled to make an alteration in the dimensions of the screws which I shall feel obliged by your introducing immediately to two thirds of your machines – the others remaining as before.

On the same day he wrote to T H Ryland, in Birmingham:

Two hampers of screws have arrived, numbers 30 and 40, which I cannot possibly receive – at all events not at the same price as the others. I might be able to use them in some unimportant situation, but certainly not on the main line.

By 11 January, the tone was becoming impatient. He wrote to Maw:

What are you doing towards our screws? I am most anxious to hear as I must extend my orders to the other manufacturers unless I have

△ Lithograph showing a pastoral scene by the Wharncliffe Viaduct at Hanwell, now in the London borough of Ealing.

immediately some satisfactory reply from you and which I should be sorry
to do after all the trouble and Expense incurred by you.
I am dear Sir,
Yours very truly
I K Brunel

The saga of the screws continues throughout the letter books. On 8 June 1838, he wrote to Ryland, ordering 'a few gross of screws', round and square-headed, 5-in (12.5-cm) long. On the same day, he sent a letter marked PRIVATE to the unfortunate Maw:

I have done all I can to meet you in this business but I cannot sacrifice the interests of the company and my credit by using inferior materials.

It may seem extraordinary that the chief engineer of such a major project should behave like a quartermaster. However, there are two factors to consider here. One is that, in the days before the telephone and e-mail, the most trivial communications had to be made by letter. (There were several collections and deliveries within a single day, so it was perfectly possible for a letter posted at breakfast time to be delivered at tea time.) The other is that such details are not as insignificant as they may seem. Brunel had learnt from Maudslay, the great mechanic who invented the screw-cutting lathe, that standardization of parts was essential for success.

Initial efforts were concentrated on the first section to Taplow, near Maidenhead, ready for a grand opening on Whit Monday, 4 June 1838. June was chosen as it coincided with three important public events: the Eton College Festival ('Founders Day'), the Royal Ascot races, and Queen Victoria's Coronation. The Eton College authorities had been vociferous in their opposition to the railway, fearing that their pupils' morals would be contaminated by easy access to London. However, although the GWR bill stipulated that no station could be built within 3 miles (5 km) of Eton, this did not prevent trains from stopping at nearby Slough, with or without a station.

Brunel had commissioned locomotives to his own specifications to be ready in time for the grand opening. According to L T C Rolt, these first locomotives represent 'the greatest and most inexplicable blunder in his engineering career'. Even Brunel's greatest supporters would concede that he was no steam engineer. He was only saved from embarrassing disaster

▷ Frontispiece from *The History and Description of the Great Western Railway* by J C Bourne. Although Brunel's name is the one most people associate with the GWR, he relied heavily on the locomotive design skills of Robert Stephenson and Daniel Gooch, who had trained in Stephenson's workshop.

by the lucky appointment of Daniel Gooch as Chief Locomotive Assistant. Gooch had worked as a draughtsman for Robert Stephenson and Co. His skill, and the availability of an engine from the workshop that had produced the *Rocket*, helped put things back on track.

In trials before the official opening day, *North Star*, the engine from Robert Stephenson's workshops, proved the most reliable of the fleet. On 31 May a group of some two hundred GWR directors rode in carriages drawn by *North Star* at an average speed of 28 m.p.h. Travelling back to

Paddington after a good lunch, one of the Bristol directors took it upon himself to walk along the roof of the train as it hurtled along at hitherto unimaginable speeds.

However, such celebrations proved premature. Although the GWR carried nearly 300,000 passengers during the month of June, share prices dropped amid rumours that all was not well. The unreliability of Brunel's original engines led to the abandonment of a timetabled service. The trains did not go as fast as had been promised, there were derailments – and there were complaints about the bumpiness of the ride.

Conventional railways – such as Robert Stephenson's London and Birmingham – had iron rails running over stone blocks, with wooden sleepers on embankments. Brunel, however, wanted to be sure his rails would be able to cope with the faster trains he planned. He bolted them to wooden longitudinal sleepers, which were in turn fixed by piles driven into the ground underneath. When experience showed that this gave too firm a ride, it was relatively easy to abandon the piles and experiment with different grades of gravel for ballast. However, the problem with the engines – and a general suspicion about the soundness of Brunel's broad gauge – was less easy to solve.

Some of the GWR directors took a ride on the London and Birmingham, and returned satisfied that it was no smoother or faster than their own railway. But this did not appease Brunel's worst critics, a group of North of England shareholders known as the 'Liverpool party'. They had vested interests in narrow-gauge lines, and wanted a greater share in the running of the GWR. They also wanted Brunel's head on a plate.

They demanded a report from an independent engineer. Robert Stephenson was approached, but he refused – a testament to the growing friendship between the two men. The two were commercial rivals – and, in fact, Stephenson had strong views on the limitations of the broad gauge – but he was not going to take the opportunity of scoring cheap points at his friend's expense.

In the University of Bristol Special Collections is a large bundle of correspondence from Stephenson to Brunel. Stephenson's handwriting was large and untidy, and many of these letters are little more than scrawled one-liners: arrangements for meeting or notes of introduction.

One letter stands out as being more carefully written than most, suggesting that it was the product of long, considered thought. In two sides of closely written script, Stephenson sets out his reasons for declining to give his opinion on the GWR:

5 August 1838

Dear Brunel,

I find it quite out of my power to pass a report on your permanent road. I have written to [name unreadable] *declining to do so. As my opinions of the system remain unchanged, you will I am sure readily see how unpleasant my position would be, if I expressed myself in an unequivocal manner in my report, and to do otherwise would be making myself ridiculous since my opinions are pretty generally known.*

To report my opinions fully therefore would do harm instead of good to the cause in which you are interested and this I am sincerely desirous of avoiding.

I have carefully considered over what I saw with you the other day on our trip to Maidenhead and I am compelled to say that my former views as to the increased width of the rails as well as the plan of laying them remain unchanged. I would put all my views on paper, but I am so pressed for time, now that our general meeting and opening throughout is approaching so near. [The London and Birmingham Railway opened on 15 September 1838.] *You are, I think, pretty well aware of my reasons, which renders such a step less necessary on my part.*

Yours very truly
Robert Stephenson

▽ Charles Babbage (1791–1871), the computer pioneer, who helped Brunel beat his critics.

He undoubtedly had strong professional grounds for declining to give the report. However, as the middle paragraph of the letter makes clear, his personal reasons were even stronger.

In his battle with the GWR directors, Brunel had another friend and ally – Charles Babbage, best known for his pioneering work with automatic calculating machines. According to Simon Schaffer, Babbage's role was crucial:

'What Babbage did was simply to travel on almost every English railway. He would count every time there was an unsatisfactory bump and his cup of tea was spilled, and he would make a note in his notebook,' he explains.

'Realizing that this was a slightly painful way of working out whether the railways were safe and stable, Babbage then designed an incredibly complicated machine which would do that task automatically.'

This involved fitting up an entire railway carriage on the GWR with a long table. Four or five mathematicians and engineers would sit round this table. At each end was a series of rollers, with eight different moving pens to make marks on the paper as it moved through. These marks would record any oscillations of the railway carriage.

'Babbage slung a whole iron plate underneath the carriage to measure the changing distance between the track and the bottom of the carriage itself – one could measure how the cars swung from side to side,' explains Simon Schaffer.

'This is really the ancestor of the modern black box on aeroplanes. It's the first automatically registered safety machine. What it meant was that when, by the spring of 1839, the shareholders and the enemies of broad gauge mobilized in force, they were met with graphs and numbers and they were blown away.

'Babbage had about 6 or 7 miles [10 or 11 km] of rolls of wallpaper, which his men pinned to the wall of the meeting room. The stockholders who'd come down from the North of England and Scotland to vote *against* Brunel at the meeting in the London Tavern were overwhelmed by this kind of evidence. They decided to vote *for* Brunel instead. One of them said that broad gauge would have disappeared from Britain had it not been for Babbage's paper.

'I think what really matters in the story of the Babbage-Brunel collaboration is that we begin to see the origins of the modern regime of things. What's absolutely crucial is that the railway system and operations research are born together. It's not just the heroic, genial construction of a romantic piece of engineering, which is so often how these systems are understood. What goes along with that is a kind of obsessively rational intelligence, which is constantly monitoring its performance. I think it's the origin of our current obsession at trying to balance the romance of travel with our passion about its safety.'

For the moment, the Great Western Railway retained its broad gauge. Most men would have rested content at fighting off such a challenge; not Brunel. As Andrew Lambert puts it:

'Every success is merely a platform for further opportunities – it's not an end state. That is why the Great Western Railway leads to the Great

△ Railway staff are dwarfed by this huge broad-gauge engine as they line up for a photographer on Paddington Station.

Western Steamship. If you go back and do more Great Western Railways, you're just a railway engineer and that's not very interesting. There's got to be more; it's got to be different.'

So it was that Brunel found his ambitions turning westwards, towards the great Atlantic Ocean.

THE GREAT WESTERN: 2

In the mid-19th century, to suggest that it was possible to steam non-stop from Bristol to New York was almost like proposing a trip to the moon.

No ship could carry enough coal to last the entire journey without refuelling, or using its sails – which left it at the mercy of the weather. Or that was what people thought. Brunel's solution to this problem was typical. Why not build it bigger?

'Brunel wasn't afraid of building a bigger ship,' explains naval historian Andrew Lambert. 'He had access to the best technology of the day. He was using the structural designs prepared by the Admiralty for their big steamships, and their even bigger sailing battleships. So he knew that the ship could be built that big; he also knew that in a ship that size he could fit the engines that he wanted to use. To do the job he had to go big – it wouldn't work on a small scale.

'People have often asked why Brunel decided to steam across the Atlantic. The more obvious question is, could he possibly have conceived of doing it any other way? Brunel is the engineer of the steam age. Sailing is old technology – no matter how well you do it, it's old-fashioned, it's yesterday. Steam is tomorrow. Brunel will steam across the Atlantic or he won't do it. What he's talking about is improved speed, improved reliability, improved communication – and that means something you can timetable.

'If you sail across the Atlantic, you never know when you're going to arrive. If you steam, you can have a timetable. You can run a service; you can leave on a set day and arrive on a set day. That's the kind of thing Brunel is after, taking the railway timetable – which has revolutionized the way people operate on land – and taking it to sea.'

The Great Western Steamship Company was established in June 1836. Its directors were to oversee the construction of the biggest steamship yet built, extending the GWR empire across the Atlantic Ocean. However, first they had to win the public relations battle.

Chief among Brunel's opponents was Dr Dionysius Lardner. As early as 1835, at a public meeting of the British Association for the Advancement of Science in Liverpool, Lardner had aired his views on the impossibility of steaming non-stop to America. In Bristol, the following year, he repeated his claims. These were the television debates of the day. They were widely publicized, and attended by a wealthy and knowledgeable audience – in other words, potential shareholders and customers of the Great Western Steamship Company. Brunel could not afford to let Lardner's message go unchallenged.

'The key arguments of the day were based on the economics of short sea steam shipping,' explains Andrew Lambert. 'Crossing the Channel in a high-powered steamship burned a lot of coal. People extrapolated from that, that it would be impossible [to cross the Atlantic], that you would burn so many tons of coal that the ship couldn't carry enough to get all the way.'

Brunel's argument was based on economies of scale. His calculations showed that while a ship's volume increases by a power of three, its surface area only grows by a power of two. The key was getting the proportions right.

'You need far more power, relatively speaking, to drive a small ship across the Atlantic, than a large one. A much bigger ship will have more room for coal and will use less of it, proportionately, so it will make the journey. Whereas a small ship, even with smaller engines, uses more fuel to move itself in proportion to its size, and is therefore not capable of crossing the Atlantic. This is where Lardner makes his mistake; he scales up directly from small steamships to large ones.'

With the help of his friend, Charles Babbage, once again Brunel won the technical argument. However, Dionysius Lardner was a powerful and persuasive speaker, and his words still left a glimmer of doubt in the minds of potential backers, which made it difficult to raise funds for the project.

Brunel was so enthusiastic that he gave his services free (as he would with all his future shipbuilding schemes). His father had worked on marine steam

△ Previous page: The scale of Brunel's shipbuilding projects ensured the presence of cheering crowds at key moments. This picture shows the SS *Great Britain* being launched in the presence of Prince Albert on 19 July 1843.

△ Print showing the *Great Western* as completed. The initial plan had been for the ship to shut down its machinery and use sail power when the weather was favourable. However, even after it was decided she would steam all the way across the Atlantic, the rig served the vital purpose of keeping her steady.

engines, so this was not entirely unexplored territory for him. Brunel also drew upon the expertise of the other members of a specially established building committee, which consisted of Brunel and three others. One was a shipbuilder, William Patterson. Another was Thomas Guppy, a wealthy engineer who ran a sugar-refining business in Bristol. (He was the GWR director who so exuberantly ran across the roof of *North Star* as it steamed back to Paddington on its first official journey.) The third was Christopher Claxton, a former naval captain who had retired on half pay to become Quay Warden of the Bristol Docks.

Captain Claxton's naval contacts enabled the team to make use of the latest ideas from the Admiralty. This included the navy's standard method of building large wooden ships using iron plates as reinforcement. Hulls were designed as rectangular structures, with the timbers running crossways. Iron plates were then bolted diagonally across them, creating a series of triangles (three-sided being a stronger geometric form than a rectangle).

'There are four major features of the *Great Western* that make it different,' says Andrew Lambert. 'It's bigger than all contemporary paddle wheel commercial steamships. The hull form, which is not Brunel's but almost certainly Patterson's [the shipbuilder], is a very fine and quite sharp hull form, which gives you much easier movement through the water. The structure of the ship is designed using this Admiralty reinforced building. Ships this big don't work without iron reinforcement.

'It has the best marine engines that money could buy from the workshops of Maudslay and Field. These were engines built by his friends to his specification. And finally, Brunel developed a new kind of rig for the ship, a rig that would assist it to cross the Atlantic, rather than try and power the ship on its own.'

Most paddle steamships at this time operated on lakes and rivers. The problem at sea was in keeping the paddle wheels engaged; one hint of rough weather, and the boat would be left with one paddle in the water, the other revolving uselessly in the air.

'Essentially the rig of the *Great Western* works by keeping the ship upright and on a steady course,' explains Andrew Lambert. 'The danger is that if the ship is allowed to roll properly, it will engage one side and another and it will literally crab its away across the Atlantic, going from side to side in almost crab-like progression, wasting much of its power going sideways.

'The role of the rig is to use the wind to hold the ship steady, so that the paddle wheels will be more effective. So what Brunel is doing is using an old technology to benefit the new technology.'

In Bristol, a workforce of some 180 labourers toiled for over a year to build the *Great Western*'s 236-ft (72-m) long hull. Modifications, including extra reinforcements to help the ship withstand North Atlantic storms, were incorporated along the way. A series of iron bolts 1½ in (4 cm) in diameter ran the whole length of her bottom frames, and her hull was sheathed in copper (against the feared *Teredo navalis* worm). Meanwhile, the machinery needed to power this vast vessel was under construction at the Lambeth workshops of Maudslay, Sons and Field. There was a 450-horsepower side-lever engine, with two cylinders and four boilers, giving a total weight of one hundred tons even when empty. Brunel – who all this time was working on the Great Western Railway – called in frequently to check on progress at both the London and the Bristol ends.

On 19 July 1837, with great ceremony, the *Great Western* was launched in Bristol. Captain Claxton cracked a demijohn of Madeira against the gilded figurehead of Neptune on her bows. Then she was safely moored off shore for the regulation banquet for directors and dignitaries, before travelling under sail to London. Here her boilers and engines were fitted, and her interior lavishly decorated.

The ship's hull was painted a dramatic black. Brunel's desire for elegant drawing room-style windows had been overruled by the more practical

△ Drawing of the *Great Western*'s main saloon, 1838. The scale of the figures makes the roof appear much higher than it actually was.

members of the building committee; but the main saloon made up for this in the florid fantasy of its decoration. There were mirror-lined recesses, entered through columns in the style of palm trees, and state cabins with intensely decorated front panels. These were the work of Edward Thomas Parris, historical painter to Queen Victoria (who had only recently ascended the throne). The March 1838 issue of *The Civil Engineer's and Architect's Journal* noted:

> *Each panel contains an admirable painting in the Watteau style, representing rural scenery, agriculture, music, the arts and sciences, interior views and landscapes. Above the doors are small panels, containing (by the same artist) beautifully pencilled paintings of Cupid, Psyche, and aerial-like figures, which considerably heighten the appearance of the saloon.*

The tone of this article is too sycophantic to be intentionally ironic. However, Cupid, Psyche, and their airy companions could literally have 'heightened the appearance' by disguising the fact that the room was only 9 ft (2.7 m) from floor to ceiling.

Such detailed descriptions provided superb public relations for the Great Western Steamship Company, which desperately needed to attract prestigious passengers.

'In the 1830s the Atlantic is *the* route,' explains Andrew Lambert. 'It connects the two most dynamic economies of the modern era; it's the one that has enough passenger and high-value goods traffic to justify what is going to be an enormously expensive undertaking. The *Great Western* is going to be the *Concorde* of her generation. She's going to be the ultimate status symbol.'

After successful engine trials, the ship set off for Bristol on the first stage of its maiden voyage to New York. Distinguished visitors – who included Marc Brunel – disembarked at Gravesend; as things turned out, this was just as well. About half an hour after this stop, there was a strong burning smell, followed by the appearance of flames at the base of the ship's funnel.

The flames turned out to be coming from the engine room, where some boiler insulation had caught fire. The Chief Engineer crawled through the dense smoke to open the feed water supply to the boilers, thus preventing an explosion; the fire was put out using a small fire engine that had been loaded on board at the last moment. However, there was one unexpected casualty. Christopher Claxton was standing below a hatchway when a heavy object crashed down upon him. After Claxton had picked himself up, he discovered the body of a man lying face down in a puddle of rising water. It was Isambard Brunel.

It was not Brunel's job to extinguish the fire, but – as always – he wanted to be in the thick of things. In his haste to help, he had tried to climb down a ladder that was already badly damaged by fire. The rungs of the ladder gave way, and Brunel fell 20 ft (6 m) to land on top of his friend. Luckily for Isambard, Claxton was bigger than him and broke his fall, thus saving Brunel's life.

'What this accident tells us about Brunel is very important,' says Andrew Lambert. 'This isn't the first time he's been injured in the line of duty, and it won't be the last. As far as Brunel's concerned, the job is the key issue, and if his personality gets in the way, so be it. He's fortunate that in falling he found a large and relatively accommodating place to land. Had he fallen onto the floor of the stokehold [the compartment containing the boilers and furnace] he might well have been killed.'

Brunel was put ashore on nearby Canvey Island to recover; luckily, his injuries were not as serious as they had at first appeared. However, his wounds kept him confined to bed at a crucial moment. Not only did he miss the rest of the *Great Western*'s maiden voyage, but this was also only two months before the opening of the first section of the Great Western Railway. Some of the difficulties that were to lead to the challenge from the Liverpool group of shareholders might have been avoided had Brunel been less distracted.

L T C Rolt suggests that, because the two strands of Brunel's story are usually told separately, people fail to make the connection between the challenge to the GWR and the challenge that arose to the Great Western Steamship Company. Both had their roots in jealousy, a resurgence of an age-old rivalry between the ports of Bristol and Liverpool.

Sceptics might scoff at the idea of a steamship providing a regular, high-speed transatlantic service – but they could not afford to ignore it. If the *Great Western*'s maiden voyage proved a success, both Liverpool and London ran the risk of being downgraded to small, provincial ports. So it was that they launched spoiler operations.

The British and American Steam Navigation Company of London, and the Transatlantic Steam Company of Liverpool each commissioned their own vessels to rival Brunel's. However, once it became clear that the new ships would not be ready in time, the two companies decided to charter steamships for the race. The London company's boat *Sirius* was only just over half the size of the *Great Western*, with nothing like the coal-carrying capacity. It had originally been designed to make the trip to Ireland, not to withstand the swells of the North Atlantic Ocean. Nevertheless, while the *Great Western* was still in the Thames, with fashionable visitors admiring its gilded saloon, the *Sirius* slipped its anchor and headed for the open sea.

Exaggerated rumours about the *Great Western*'s safety went before it to Bristol; by the time the ship reached that port, fifty passengers had asked for their money back. So it was with a depleted cargo of only seven passengers – one woman and six men – that the *Great Western* finally set out for New York on Sunday 8 April. The *Sirius*, meanwhile, had left Cork four days earlier.

'The race to be the first to cross the Atlantic isn't a hundred-metres dash where they start off side by side – they start off at different times and at different places,' explains Andrew Lambert.

'The *Sirius* has to leave from the south coast of Ireland, because she carries so little fuel, she won't make it if she leaves from England. The *Great Western* sets off later and from further back, but she travels faster because she's bigger and more powerful. So she's going to make the first really efficient crossing. The *Sirius* is just going to make it; the *Great Western* is going to steam across triumphantly.'

Whatever the qualities of the ship, a transatlantic crossing under steam was hardly going to be a ride on a punt. After the *Great Western*'s first accident, several stokers quietly slipped overboard and disappeared in a boat. During the passage across the Atlantic, their crewmates must have wished they had followed them. The first-class passengers might have had good quality wine and fine food to distract them (especially since on this journey they were outnumbered by stewards and waiters). But the men who toiled in the bowels of the ship had little to alleviate the hard physical labour.

Keeping the furnaces red-hot, thereby maintaining boiler pressure, would have been a sweaty, exhausting, round-the-clock job at the best of times. Coal had to be shovelled in at a rate of almost a hundredweight (50 kg) a minute. However, as the journey progressed, the coal would have become more and more inaccessible. Once the stores nearest to the furnaces had been used up, the coal 'trimmers' would have to transport wheelbarrow loads from ever more distant bunkers. This would be hard enough on dry land. In a rolling, pitching, ship, with the barrow alternately running away and then having to be pushed up a slope, it must have been a nightmare.

On *Sirius*, the nightmare had an extra dimension as the ship started to run out of fuel. Myth has it that the crew fed everything combustible into the boiler, including a mast and a child's doll. In reality it was four barrels of resin – originally carried as cargo – that saved the day. This highly combustible substance was mixed with clinker in a last desperate effort to keep the engines going. It worked, and the ship steamed into New York after nineteen days at sea.

However, although *Sirius* won the sprint, the *Great Western* won the marathon. She arrived in New York only a few hours afterwards, and with 200 tons of coal to spare. As Andrew Lambert puts it: 'The significance of being the first steamship to cross the Atlantic is that it's the one that gets noticed – it's the *Guinness Book of Records* mentality. It doesn't actually matter. The one that matters is the one that makes a profit and carries on running – and that isn't *Sirius*, it's the *Great Western*. The *Great Western* is the ancestor of modern intercontinental mechanically powered travel, and the way we move today is merely a development of what Brunel does with the *Great Western*.'

When the *Great Western* made the return journey to Bristol, she carried sixty-eight passengers – a clear vote of confidence in this new mode of transport. She had not completed her second voyage before Brunel started

△ Belching black smoke, the *Great Western* dwarfs the *Sirius* at dock in New York. Picture from a French book, *Merveilles de la Science*.

planning a bigger and better sister ship. The company laid in stocks of timber and prepared to start work. However, restless as ever, Brunel was not content simply to repeat his first success. If the *Great Western* had shown the strength of iron as reinforcement, why not abandon timber altogether and build the new ship entirely of iron?

The first iron-floating opportunity was created by an 18th-century Midlands iron-founder called John Wilkinson. His friends called him 'Iron Mad' Wilkinson, because he made everything of iron – even a coffin. When he made an iron boat, sceptics said it would sink. However, the joints were well sealed with tar and pitch and it floated successfully.

'Wilkinson's iron boat was certainly noted, but I suspect that most people dismissed it as just another demonstration by Mr Wilkinson of his favourite subject,' says Andrew Lambert. 'If he'd made an iron moon rocket, they would have been equally unimpressed. The man and the material were already a standing joke, so the fact that it worked was forgotten in the process.'

However, by the end of the Napoleonic Wars, in 1815, shipbuilding timber of the right quality and the right price was becoming hard to find. In its place, high-quality wrought-iron plate, at an affordable price, started to come onto the market. Engineers were already creating massive iron constructions on land. It had been proved that iron could float, so it was not going to be long before someone made the intellectual leap. An iron boat, the *Vulcan*, was built in Scotland in 1817, and from then on it was only a question of time.

Many iron ships had been built by the time Brunel entered the scene. However, these could only be used within sight of land, as the iron prevented compasses from working properly. Various Heath-Robinson solutions had been tried. These included putting the compass at the top of the mast, or trailing it in a copper-plated boat, with a boy in the boat to relay the information by means of semaphore. None were ideal, as the ship's captain needed the information at hand, by the steering position. Before anyone could think of crossing an ocean in an iron ship, they had to have a reliable method of navigation.

The problem of the iron ship and the compass was a simple one. A compass relies on finding magnetic north, and an iron ship is a large magnetic structure. It will deflect the action of the compass and find north all over the place, but almost certainly not where it is meant to be.

In the mid-1830s, the Astronomer Royal, Sir George Airy, developed a system of iron weights and magnets set alongside the compass to compensate for this effect. It was not wholly reliable, but it was good enough most of the time. This allowed the iron ship to escape from coasts and rivers and to move out onto the open sea.

The first serious iron sea-going ship was a paddle-wheel vessel called *The Rainbow*. When this came to Bristol, Brunel was impressed. What struck him was not so much the ship itself, but the fact that she could venture out to sea without getting lost. If an iron ship could travel out of sight of land, she could cross the Atlantic. So it was *The Rainbow* that kick-started the shift from wooden paddle-wheel vessel to iron paddle-wheel vessel.

Building a timber sea-going ship required massively thick chunks of wood. If the wood started to rot, the whole framework had to be dismantled to replace it. Iron was more durable, despite its tendency to rust – and the technology developed for railway steam engines meant that high-quality wrought-iron plate was now relatively cheap, and easily available. The wooden ship was an art form, built by skilled craftsmen; the iron ship was a precision piece of engineering.

Just as Brunel was able to use iron because of processes originally designed for the railway industry, so his shipbuilding work on what was to become the SS *Great Britain* fed into other areas of engineering. The best example of this is the development of the steam hammer. The ship as

△ This portrait of James Nasmyth formed the frontispiece to his autobiography, ghost-written by Samuel Smiles.

originally planned included designs for the biggest ever set of paddle-wheel engines. This in turn meant that the shaft, which connected the two wheels and carried the drive, would be the biggest and heaviest iron shaft ever made. No existing foundry technology could make that shaft.

Cue James Nasmyth, the engineer and machine-tool manufacturer. Born in Edinburgh, Nasmyth was the classic example of the intrepid Victorian entrepreneur. He left school at twelve, and, by the age of seventeen was a skilled metal worker. He raised the money to attend university classes by turning his bedroom into a mini foundry, making brass castings in the fireplace, and selling models of steam engines at £10 a time.

In his autobiography, published in 1883, Nasmyth recalled:

My brass foundry was right over my father's bedroom. He had forbidden me to work late at night, as I did occasionally on the sly. Sometimes when I ought to have been asleep I was detected by the sound of the ramming in of the sand of the moulding boxes. On such occasions my father let me know that I was disobeying his orders by rapping on the ceiling of his bedroom with a slight wooden rod of ten feet that he kept for measuring purposes. But I got over that difficulty by placing a bit of old carpet under my moulding boxes as a non-conductor of sound, so that no ramming could afterwards be heard. My dear mother also was afraid that I should damage my health by working so continuously. She would come into the workroom late in the evening, when I was working at the lathe or the vice, and say, 'Ye'll kill yersself, laddie, by working so hard and so late.' Yet she took a great pride in seeing me so busy and so happy.

This episode ranks alongside James Watt and his steam kettle as a typical example of Victorian myth-making at work. Nasmyth's autobiography was supposed to have been 'edited' by Samuel Smiles, the author of *Self-Help* and *Lives of the Engineers*. In fact, Smiles actually wrote it, using Nasmyth's notes and diaries as the principal sources.

Nasmyth's (or rather, Smiles's) account of his invention of the steam hammer should be set within this context. For in a chapter entitled, bizarrely, 'My Marriage – The Steam Hammer', he describes how he 'sketched out' the idea for his new device in little more than half an hour. Included in the autobiography is an engraving of the original page torn from his sketchbook. It is hard to believe that such immaculately drawn illustrations, with detailed calculations to accompany them, really took so little time.

Whatever the truth of its conception, Nasmyth's steam hammer was an ingenious contraption. The problem was simple. Mechanized versions of the traditional blacksmith's hammer already existed – but they could not be lifted high enough to come down with sufficient force on a rod of iron 30 in (76 cm) thick. Nasmyth's machine used steam to lift a large iron block and make it crash down when released. A machine attendant controlled the force with which the weight descended on the anvil by opening or closing steam valves.

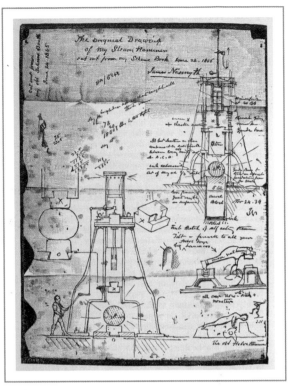

△ Nasmyth's first sketch for the steam hammer, which took the human effort out of the process of forging iron.

The steam hammer was destined to become an icon of the Industrial Revolution. However, the giant paddle shaft for Brunel's latest ship was never forged. No sooner had Nasmyth sent his designs to the Great Western Steamship Company than Brunel was persuading the directors that paddle wheels were yesterday's technology. The future lay in screw propellers.

Poor Francis Humphrys, the young Chief Engineer on the project, had been hoping to make his name with the giant paddle engines for the new ship. He had produced elaborate designs, and the engines were well under construction. However, seemingly on a whim, he was ordered to abandon work on these and start all over again. As Nasmyth put it:

Mr Humphries [sic] was a man of the most sensitive and sanguine constitution of mind. The labour and the anxiety which he had already undergone, and perhaps the disappointment of his hopes, proved too much for him; and a brain fever carried him off after a few days' illness.

△ Nasmyth's steam hammer was a machine that combined great strength with precision. His favourite party trick was to place an egg in a glass on the anvil, and use the hammer to crack the top of the egg.

The company had lost their Chief Engineer; they had also lost the money they had invested in the engines that were now to be abandoned like so much scrap iron. However, Brunel was adamant that the screw propeller was the way forward.

His conviction was based on a 237-ton demonstration ship called the *Archimedes*, which had called into Bristol on a round-Britain tour to promote the virtues of screw propulsion. The *Archimedes* was fitted with a screw propeller low down in the water, at the back of the ship. This meant that she avoided all the problems of a paddle-wheel vessel – no more side-to-side motion, no more variation in efficiency as the ship's cargo was emptied.

Thomas Guppy, of the building committee, took a trip round the coast to Liverpool and returned with a highly favourable report on her performance in rough seas.

'Immediately, Brunel sees that he's got a new kind of ship. Iron hull, screw propeller... he's solved all of the problems that the *Great Western* had merely skirted round,' says Andrew Lambert.

'This is a modern ship. And the new ship, to celebrate its size and its impressive nature, is given the name the *Great Britain*, to celebrate a national demonstration of engineering excellence. And also to say the Great Western Steamship Company is a national enterprise; this isn't just Bristol, this is the country.'

The screw propeller was not a new invention. In Ancient Greece, Archimedes had used a screw to lift water up a hill, hence the name given to the demonstration ship. In the 19th century, various engineers had tried fitting screw propellers on boats. They failed because they put the propellers in the wrong place, and did not use enough power to drive them. As so often happened, it was not necessarily the man who had the first idea who succeeded, but the one who believed in the discovery so much he was willing

△ The *Archimedes* (right) and a paddle wheel steamer in a storm. This promotional print was designed to show the superiority of screw-driven ships in bad weather.

to risk everything to make it happen. That man was Francis Pettit Smith.

Smith, a sheep farmer from Kent, had experimented with model boats ever since he was a child. Having no engineering training, he progressed by trial and error – and his greatest breakthrough came as a result of an accident.

'The first experiment was with a rowing boat on the Regent Canal in London,' says Andrew Lambert. 'He had a propeller with two full turns of screw on it, and as he was driving up the channel of the canal, one of those turns was knocked off by a piece of floating rubbish and the boat then went faster.

'So he learned a fundamental engineering principle by accident. The engineering principle he learned was that a multi-turn Archimedean Screw is a water-lift. It's not the most efficient way of putting power into the water. And for the next five years it will gradually go from a long structure to a very narrow, flat structure. What Brunel creates in the mid-1840s is completely different to what Pettit Smith starts off with.'

The Great Western Steamship Company chartered the *Archimedes* for a six-month-long experiment. Brunel also carried out screw propeller trials for the Admiralty, culminating in a typical piece of showmanship. This was a tug of war between a paddleboat, *Alecto,* and a ship armed with a screw propeller, the *Rattler.* Both ships were identical, apart from their means of propulsion, and the *Rattler* towed her rival away at a rate of 2.8 knots.

Brunel's publicity stunt convinced even the most conservative members of the Admiralty that the future belonged to the screw propeller. However, behind the razzmatazz was a series of very practical experiments. These led to the gradual change in shape of the screw propeller to something considerably shorter and flatter than Pettit Smith had envisaged.

Smith's patent was not for the propeller itself, but for its placing, as Andrew Lambert explains:

'The great success of Pettit Smith's propeller location is that it's right where the water flows past the hull of the ship. And on a ship like the *Great Britain*, with very fine lines towards the stern, the water comes past unbroken, it's smooth, and the propeller then bites into that water to drive the ship forward. When the propeller is doing that, it's putting a lot of pressure on the rudder. This means the propeller can actually steer the ship.

'Sailing ships are not steered using the rudder, they're steered using the sails, and the rudder is just a little trim-flap that helps you to tidy up around the margins. On the *Great Britain*, a pure steamship, you need the rudder to control the direction the ship is going in. So what Brunel is

▽ Following pages:
Tug of war between
Rattler (left) and *Alecto.*
April 1845.

doing is taking Francis Pettit Smith's idea about where the propeller goes, other people's ideas about iron ships, and making them all come together in a package with this brilliant piece of innovative design, the balanced rudder. That's his signature, and also the final piece in the jigsaw. Without this balanced rudder, the *Great Britain* would not have gone anywhere except round the corner.'

On 19 July 1843, the *Great Britain* was named and launched by Prince Albert, who had travelled down to Bristol by special train from Paddington. There was a banquet, a bottle-crashing ceremony, then the sluices of the dry dock opened, and the ship was towed out into the floating harbour.

However, the celebrations proved somewhat premature. The *Great Britain* had been built too wide for the narrow locks that she had to pass through, and she was held up by lengthy negotiations with the Bristol Dock authorities. Even after one lock had been temporarily widened, she nearly stuck fast in another. Brunel told the story in a letter of apology to the directors of the South Wales Railway, whom he had been scheduled to meet the next day:

11 December 1844

We have had an unexpected difficulty with the Great Britain. *She stuck in the lock; we did get her back. I have been hard at work all day altering the masonry of the lock. Tonight, our last tide, we have succeeded in getting her through; but, being dark, we have been obliged to ground her outside, and I confess I cannot leave her till I see her afloat again, and all clear of her difficulties. I have, as you will admit, much at stake here and I am too anxious to leave her.*

When she finally reached London, the *Great Britain* had a five-month stay in the Thames. Queen Victoria paid a visit, and Brunel explained his engines to her using a working model, as it was assumed that it would be beneath her dignity to descend into the engine room itself. Thousands of visitors came to admire the ship's elaborate state rooms, adorned by 1,200 yards (1,100 m) of specially woven scarlet and purple Brussels carpet.

After her maiden voyage from Liverpool to New York in July 1845, the *Great Britain* made four relatively uneventful round trips. There were modifications to her boilers to make for more effective steaming, but nothing major. However, disaster struck in September 1846. A few hours after her departure from Liverpool, with a record 180 passengers on board, she ran aground. It was the middle of the night, and the captain thought they were near the Isle of Man; in fact, as daylight broke, he discovered they were in Dundrum Bay, County Down. (It later transpired that the ship had been using faulty charts, which did not show a vital beacon.)

The passengers were taken to safety by a procession of Irish carts, and the sailors tried to re-float the ship by lightening the cargo. This failed. Christopher Claxton arrived from England and tried to take advantage of a high spring tide; however a violent gale sprang up and he was forced to abandon the attempt. Instead, he used the ship's sails to drive her further up the sand, where she would be slightly more sheltered.

When Brunel finally visited two months later, he was furious to find his precious ship lying beached and exposed to the elements. He wrote to Christopher Claxton:

▽ SS *Great Britain* in rough seas off Lundy. Engraving by unnamed artist in the *Illustrated London News*, 1 February 1845.

I have returned from Dundrum with very mixed feelings of satisfaction and pain, almost amounting to anger, with whom I don't know. I was delighted to find our fine ship almost as sound as the day she was launched, and ten times stronger in character. I was grieved to see her lying unprotected, deserted, and abandoned by all those who ought to know her value and ought to have protected her. The result, whoever is to blame, is that the finest ship in the world, in excellent condition such that £4,000 or £5,000 would repair all the damage done, has been left lying like a useless saucepan, kicking

*about on the most exposed shore you can imagine, with no more effort or
skill applied to protect the property than the said saucepan would have
received on the beach at Brighton.*

With winter storms already threatening, there was no hope of floating her off
before the spring, so Brunel devised a protective cover. This was a massive
breakwater of faggots, lashed together and skewered with iron rods, and
weighted down with iron and sandbags.

Brunel's homespun barricade worked, and when the salvage operations
finally began, the *Great Britain* was still in good shape. She had weathered
storms that would have reduced a wooden ship to splinters. Under
instructions from a Scottish salvage expert, all her stores, furniture and

△ SS *Great Britain* with
protective barricade at
Dundrum Bay – winter
1846/7. The ship
survived, but the
company did not.

fittings were removed so that she was half the weight she had been when she had gone aground. Teams of Irish navvies built trenches beneath her, and repairs to the bottom of her hull were carried out, where it had been holed in two places.

Eventually, she was raised by attaching great boxes of sand, each weighing thirty tons, to the forward end of the ship. When full, these boxes tended to move the ship. By the end of August 1847, the *Great Britain* was back in Liverpool. However, although the ship was saved, the Great Western Steamship Company was not.

Earlier in the year, the company had been forced to sell the *Great Western* to fund the salvage operation on her sister ship. This left them with no regular source of income. They had been underinsured, and once the *Great Britain* was repaired, they were forced to let her go, too. In September 1847, the ship was put up for auction, and the company was wound up.

Disillusioned by this sequence of events, Brunel abandoned his shipbuilding ventures for the next few years. The last great project of his career was to be a final attempt at creating a world-beating leviathan; meanwhile, he returned his attention to the railways.

THE GREATEST BLUNDER

Brunel's supporters were used to his wild, unpredictable enthusiasms. However, even their loyalty was tested by his flirtation with the railway system dismissed by George Stephenson as a 'rope of wind'.

The idea behind the atmospheric railway was to have trains without engines, which could be whisked up and down hills using the power of air pressure. The carriages would be connected to a piston gliding through a tube between the rails. A series of stationary steam engines would suck the air out of the tube ahead of the piston and thus pull the train along.

Using pneumatic trains would mean passengers could avoid getting covered in smuts from locomotives. Pneumatic trains would be cleaner, quieter and more efficient – at least that was the idea. However, sceptics pointed out that they still needed steam engines – albeit stationary ones – which made them little better than old-fashioned rope-hauled systems.

When Brunel was made Chief Engineer for the South Devon Railway, extending the GWR empire westwards from Exeter to Plymouth, he saw the atmospheric system as a solution to the increasingly difficult terrain he had to cross. With no locomotives to worry about, he could take his trains up much steeper slopes.

Daniel Gooch, the locomotive engineer whose steam engines had done so much to improve the reliability of the GWR, was understandably unimpressed. In September 1844, he went with Brunel and other engineers to inspect the atmospheric railway that ran in Ireland, between Kingstown (now Dun Laoghaire) and Dalkey. Its initial success, linked with the novelty of the idea, made it superficially attractive.

Brunel was smitten – but Gooch calculated that he could do the work much more cheaply with locomotives. In his memoirs, he recalled:

> *I could not understand how Mr Brunel could be so misled... he had so much faith in his being able to improve it that he shut his eyes to the consequences of failure.*

A patent for transporting goods and mail at high speed through a pneumatic iron tube had first been taken out by a Soho scales manufacturer in 1799. Various attempts had been made to apply this principle to railways, but it proved difficult to maintain an airtight seal. Eventually, the gas engineer Samuel Clegg, together with the brothers Jacob and Joseph Samuda, managed to produce a working model. They laid out a ½-mile (0.8-km) test track at Wormwood Scrubs, using old lines from the Liverpool and Manchester Railway. Trains ran for two hours every Monday and Thursday afternoon, and engineers, thrill-seekers and the merely curious went along for a ride.

The engineering world divided into those who considered – like George Stephenson – that it was a 'great humbug', and those who saw atmospheric railways as the future. During the sixteen years of atmospheric frenzy in Britain, four such railways, totalling some 30 miles (48 km) of single track were opened. The last one closed in 1860.

On paper, the proposition was attractive. A pamphlet entitled Clegg's Patent Atmospheric Railway suggested that the building and working costs would be around a quarter that of a conventional steam railway. The rails would be lighter, hills would no longer be a problem and there would be less need for cuttings and embankments.

However, there were snags. One was highlighted when an Irish engineering student named Frank Ebrington accidentally created a speed record. This unfortunate lad had been sitting waiting to make a test run on the Dalkey to Kingstown railway, when his carriage was suddenly sucked away along the track without its train. Someone had forgotten to couple it up and he shot off at an average speed of 84 m.p.h.

△ Previous page: Bristol Goods Shed, 1842, by J C Bourne. The Great Western Railway had made Brunel's name. However, his dalliance with atmospheric railways tested the loyalty of even his staunchest supporters.

△ Proposed atmospheric station for the Epsom extension of the London and Croydon Railway in 1845. This strange mixture of architectural styles, complete with gothic bell tower, was no doubt intended to inspire confidence in potential investors. However, enthusiasm for the atmospheric experiment was short-lived.

Even under normal circumstances, those on the train were at the mercy of the engine men in a pumping station often several miles up the track. The train driver had a gauge to show what vacuum he had in the tube. He could slow the train with his brake, but could do nothing to put on speed. It was not unusual for the train to stop short and have to be pushed into place, or to overshoot the platform. Wooden platform extensions were built to cope with this eventuality.

Despite all this, Brunel was so sold on the idea that he wanted to introduce atmospheric working not only in South Devon but also on a new line proposed between Newcastle and Berwick. This was deep into Stephenson territory. Legend has it that when Brunel visited Newcastle, George Stephenson seized him by the collar and asked him what business he had north of the Tyne. The tone may have been playful, but – since both George and Robert Stephenson were so set against atmospheric railways – the intrusion must have been doubly unwelcome.

Robert Stephenson had reported on the atmospheric system for the directors of the Chester and Holyhead Railway, of which he was engineer. He came out firmly against it, giving the example of his London and Birmingham Railway. If this had been run on the atmospheric principle, it would have needed a chain of 38 pumping stations – and everything would come to a grinding halt if just one broke down. He was also sceptical about the great economy promised by the atmospheric's supporters, and pointed out the number of man hours spent on maintaining the Dublin railway.

His comments were prophetic. The first stretch of the South Devon Railway opened with conventional steam locomotives, instead of the whizzy new system, as the pumping stations could not produce enough power. By the autumn of 1847, atmospheric trains were eventually working the single-line track from Exeter to Teignmouth. But then another problem surfaced – the English weather.

The reliability of the system depended on maintaining an airtight seal around the piston. This was connected to the train above through a slit in the top of the pipe. A series of leather flaps covered this slit. However, the leather froze and leaked rain in winter, and cracked in summer. Air escaping from the damaged sections caused the pumping engines to work overtime, and an army of men had to be constantly employed rubbing sealant on the leather. On one occasion, a damaged valve caused a train to get stuck near a tunnel. The train staff had to get out and push, plugging the hole with a railwayman's cape.

Once people started to lose confidence in the idea of atmospheric working, it became a scapegoat for everything. As the South Devon used both the conventional and atmospheric systems together, this meant that the atmospheric never got a fair trial. Brunel's son, in his biography of his father, noted:

The Atmospheric System was vaguely credited with every delay which a train had experienced in any part of its journey; though, in point of fact, a large proportion of these delays was really chargeable to that part of the journey which was performed with locomotives. It often happened that time thus lost was made up on the Atmospheric part of the line, as is shown by a record of the working, which is still extant. In the week, 20–25 September 1847, it appears that the Atmospheric trains are chargeable with a delay of 28 minutes in all; while delays due to the late arrival of the locomotive trains, amounting in all to 62 minutes, were made up by the extra speed attainable on the Atmospheric part of the line.

In his report to the directors in February 1848, Brunel was still optimistic. The only real difficulty he admitted to was that of communications. A telegraph system was being introduced, but until this was established, those manning the boilers in pumping stations along the line were never sure when a train was about to come. This meant they had to keep their fires red hot almost all the time, just in case they needed to get up a head of steam.

However, by the summer, this problem had paled into insignificance beside the difficulties with the leather flaps. Faced with the prospect of having to renew the whole length from Exeter to Newton Abbot, at a cost of £25,000, Brunel decided he had no choice. He wrote to the directors:

> *The cost of construction has far exceeded our expectations, and the difficulties of working a system so totally different from that to which everybody, traveller as well as workmen, is accustomed, have proved too great; and therefore, although, no doubt, after some further trial, great reductions may be effected in the cost of working the portion now laid, I cannot anticipate the possibility of any inducement to continue the system beyond Newton.*

He went on to say that, unless the patent-holders would be prepared to renew the valve flaps at their own expense, he did not think the directors would be justified in spending any more money. In other words – since the patent-holders were extremely unlikely to come up with £25,000 – they should give it all up as a bad job.

L T C Rolt sees this as a courageous decision. As he put it: 'Squarely he faced the fact that he had been responsible for the most costly failure in the history of engineering at that time. There could be no shirking the issue; the slate must be wiped clean.' It could be argued that it is always easy to be courageous with someone else's money. However, in fairness to Brunel, it should be pointed out that he was not the only one to be seduced by the atmospheric promise. William Cubitt, a prominent canal and railway engineer, used the system from London to Croydon with no more success. And the shareholders of both their companies were equally enthusiastic until things started to go wrong.

The Editor of the *Railway Times* wrote this exasperated comment on Brunel:

> *We do not take him for either a rogue or a fool but an enthusiast, blinded by the light of his own genius, an engineering knight-errant, always on the lookout for magic caves to be penetrated and enchanted rivers to be crossed,*

MANNERS·AND·CVSTOMS·OF·Yᵉ·ENGLYSHE·IN·1849. Nᵒ·19.

·A·RAYLWAVE·MEETYNGE·   EMOTYON·OF·Yᵉ·Shareholderes·AT·Yᵉ·ANNOVNCEMENTE·OF·A·DIVIDENDE·OF·2ᵈ·½

never so happy as when engaged 'regardless of cost' in conquering some, to ordinary mortals, impossibility.

According to railway historian Adrian Vaughan, the whole atmospheric episode is a supreme example of Brunel playing the 'engineering knight-errant'. The system might have been practical for a short length of track in Ireland, but it was hopeless for anything more ambitious.

'The atmospheric railway requires a 15-in [38-cm] diameter cast-iron tube laid all along the middle of the track,' he explains. 'So $7\frac{1}{2}$ in [19 cm] of cast-iron tube is sticking out of the ground, and the thing travels along the slot at the top. What happens when you want to turn across the tracks? You can't. So the only way you can have a set of points is to stop the tube. If you stop the tube you've got no power.

◁ The railways offered *Punch* cartoonist Richard Doyle much comic potential. Small shareholders were attracted by the promise of enormous profit – but they paid the price when things went wrong. Brunel's flirtation with atmospheric railways proved the most expensive failure in engineering history of his time.

'I've got paintings that were made by the South Devon Railway Company in 1846, and pictures of Newton Abbot, and the only track to have the tube in it is the running line. All the sidings have no power, so you've either got to have a steam engine to push your coaches around, or you push them with your shoulder. You can only go one direction with an atmospheric railway. You can't shunt, can't come backwards.

'Every few miles you're going to have a pumping station, pumping the air out of the tube; the cost of a pumping station would have been enough to buy a fleet of steam engines. Brunel wouldn't listen. It's not that it didn't work. It did work – it just wasn't practical.

'On the final meeting where Brunel said, "We've got to abandon this", one of the shareholders stood up and said that Mr Brunel had lost £400,000 of money contributed by many, many small investors that they could ill afford to lose. It wasn't right that Mr Brunel should have played with the property of the shareholders when engineers at least as eminent as himself had told him that it couldn't work.

'What Brunel did by way of penance, was to delay being paid his salary as engineer of the railway until he got it open. So he went without his pay for two years, from 1847 to 1849 [when the South Devon opened with locomotive working], and then he was paid.'

The miles of cast-iron tubing were sold for scrap; meanwhile the army of leather valve-menders had to find other jobs. The Teignmouth correspondent of the *Western Times* wrote their lament, in the style of Othello's farewell speech:

> *Farewell the tranquil mind – farewell content!*
> *Farewell, ye pistons, pipes, and valves, and all!*
> *Farewell extensive sheds, and the fat grease*
> *That makes the valves adhesive. All farewell!*
> *Adieu, great air-exhausting engine-houses!*
> *The Atmospheric occupation's gone.*

In his *Memoirs*, Daniel Gooch provided a less poetic epitaph, noting: 'This is certainly the greatest blunder that has been made in railways.'

Meanwhile, Brunel had been under siege over another issue: the broad gauge. All Brunel's railways in England – the South Devon included – had been constructed to a 7-ft (2.1-m) width. However, as every other railway engineer in the rapidly expanding network was using Stephenson's narrow

◁ Front view of the GWR broad-gauge engine *Balaklava*, on a mixed-gauge track.

gauge, this was causing increasing aggravation. No one was exempt; Queen Victoria had to change gauges twice on her journey from Windsor to Balmoral.

A mixed-gauge track, with an extra length of rail laid inside the broad gauge, enabled both types of trains to use the same stations. However, transfers involved the wholesale emptying of one train and the filling of another. Those who supported the narrow gauge against the broad gauge had a vested interest in making sure things were as chaotic as possible. The most publicized example was at Gloucester, where the Midland Railway met an extension of the Great Western Railway. The *Illustrated London News* wrote:

> *It is found at Gloucester that, to trans-ship the contents of one wagon full of miscellaneous merchandise to another, from one Gauge to another, takes*

*about an hour, with all the force of porters you can put to work on it…
In the hurry the bricks are miscounted, the slates chipped at the edges, the
cheeses cracked, the ripe fruit and vegetables crushed and spoiled; the chairs,
furniture, oil cakes, cast-iron pots, grates, and ovens all more or less broken:
the coals turned into slack [coal dust], the salt short of weight, sundry
bottles of wine deficient, and the fish too late for market. Whereas, if there
had not been any interruption of Gauge, the whole train would in all
probability have been at its destination long before the transfer of the last
article, and without any damage or delay.*

▽ 'Break of Gauge at
Gloucestershire', 6 June
1846. Cartoon from the
Illustrated London News,
with a comic version of
the chaos caused by
transferring from one
gauge
to another.

Parliament's response was – as usual – to set up a Royal Commission. A total
of forty-four witnesses spoke for the narrow-gauge interests, with only four
giving evidence for the broad gauge. These included Brunel and Gooch, a
pair dismissed by pamphleteers as Don Quixote and his Sancho Panza. Since
by this stage of the proceedings there were 1,900 miles (3,000 km) of narrow
gauge, as opposed to 270 miles (430 km) of broad gauge, it did seem as if
Brunel was tilting at windmills. However, the fact that the commission was
set up at all is a testament to the power of his vision.

Transcripts of the testimony highlight the fundamental difference in
temperament between Brunel and Robert Stephenson. Stephenson made it

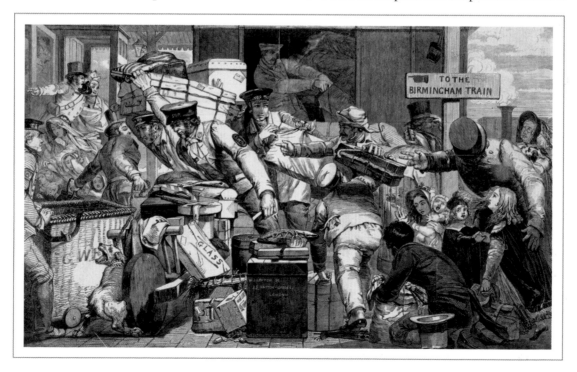

clear that he used the narrow gauge simply to conform to the width already used by his father. By contrast, Brunel proved to be his usual maverick self, as the following extract illustrates:

> *Q. Had you, before you took the management of the Great Western Railway, any employment in railway matters?*
>
> *A. No.*
>
> *Q. Having seen the working of other railways, and of the Great Western since its entire opening, are you inclined to think it was an injudicious arrangement to alter the Gauge to 7 feet, or that a less difference would have been better?*
>
> *A. To answer that, as I will endeavour to do, with candour, I incur the risk, I am afraid, of being accused of adopting wild notions; I should rather it be **above** than under 7 feet now if I had to reconstruct the lines.*

Never shy of a publicity stunt, Brunel offered to take the duel of the gauges onto the railway tracks. Two courses were chosen: Paddington to Didcot and York to Darlington. It was like the Rainhill Trials all over again, only this time a whole railway network was at stake.

The gauge opponents set about choosing their weapons. Daniel Gooch chose *Ixion*, a 7-ft (2.1-m) locomotive built in 1841. The narrow-gauge party selected a North Midland engine called *Stephenson*, and a locomotive so new it was only referred to as 'Engine A'. Brunel's plan had been that all the commissioners should ride on the footplate; in the event, only George Airy, the Astronomer Royal, took up the challenge.

In mid-December 1845, *Ixion* made three round trips between Paddington and Didcot. She reached a top speed of 60 m.p.h., with a load of eighty tons, the feed water in the tender being pre-heated for maximum efficiency. 'Engine A' not only had pre-heated water, but had a portable boiler waiting at Darlington to give it a flying start for the return trip. Even so, she only achieved a maximum speed of $53\frac{3}{4}$ m.p.h. with a load of fifty tons. The *Stephenson's* challenge came to an embarrassing end, when she ran off the rails after only 22 miles (35 km). Luckily, the distinguished Astronomer Royal was not on board at the time.

After such a success, the Commissioners' report, however inevitable, was a disappointment to Brunel. Although they acknowledged that the broad-gauge trains could go faster, they concluded that – for the sake of uniformity – the narrow gauge should become the national standard. Brunel and his supporters campaigned vigorously to make sure that this did not become law. In the end, the Gauge Act of 1846 produced an awkward compromise. All new lines had to conform to the standard gauge, unless they were extensions of existing broad-gauge networks. So it was that the broad gauge lingered on, like the bleached skeleton of a Megalosaurus.

Had matters come to a head at an earlier stage, we might still be riding around on a 7-ft (2.1-m) gauge. However, too many people had too much at stake; it was a simple question of economics. In his diaries, Daniel Gooch conceded that he had been fighting a lost cause – but felt that some good had come out of it. He wrote:

> *Were the whole question now open to be decided, the broad gauge is safer, cheaper, more comfortable, and attains a much higher speed than the narrow, and would be the best for the national gauge. But as the proportion of broad to narrow is so small, there is no doubt the country must submit to a gradual displacement of the broad, and the day will come when it will cease. The fight has been of great benefit to the public; it has pricked on all parties to exertion; the competition of the gauges has introduced high speeds and great improvements in the engines, and was of great practical use to all those who were actively mixed up in the contest, as they were forced to think and experiment. It was not allowed to them to rest quietly on speeds of twenty to thirty miles per hour. I know it was of great value to me by the practical information I obtained in investigations.*

In his biography of George Stephenson, Samuel Smiles conceded that the gauge wars had been a stimulus to invention. However, he made it clear that he thought Brunel headstrong and impractical at best:

> *Mr Brunel had always an aversion to follow any man's lead; and that another engineer had fixed the gauge of a railway, or built a bridge, or designed an engine, in one way, was of itself often a sufficient reason with him for adopting an altogether different course. Robert Stephenson, on his part, though less bold, was more practical, preferring to follow the old routes, and to tread in the safe steps of his father.*
>
> *Mr Brunel, however, determined that the Great Western should be a giant's road, and that travelling should be conducted upon it at double speed.*

His ambition was to make the best road that imagination could devise; whereas the main object of the Stephensons, both father and son, was to make a road that would pay.

Despite the fact that Brunel and Stephenson were directly opposed to each other over the gauge question, they remained friends. On 5 May 1846, Brunel wrote in his diary:

I have just returned from spending an evening with Robert Stephenson. It is very delightful in the midst of our incessant personal professional contests, carried to the extreme limits of opposition, to meet thus on a perfectly friendly footing and discuss engineering points as if we were united. Stephenson is decidedly the only man in the profession that I feel disposed to meet as my equal, or superior, perhaps, on such subjects. He has a truly mechanical head and it is singular that we never differ in the end when we thus meet, although always differing apparently as black from white in our public discussion.

It was inevitable that the two foremost railway engineers in the country should increasingly find themselves in competition as Britain went railway mad. During the period from 1845 to 1848, Parliament passed a total of 650 acts, authorizing the construction of 9,000 miles (14,500 km) of track. Each act took up hours of parliamentary time, and when Robert Stephenson became an MP in 1847, he found himself in growing demand. As Samuel Smiles put it:

Much of his labour was heavy hackwork of a very uninteresting character. During the sittings of the committees of Parliament, almost every moment of his time was occupied in consultations, and in preparing evidence or in giving it. The crowded, low-roofed committee rooms of the old Houses of Parliament were altogether inadequate to accommodate the rush of perspiring projectors of bills, and even the lobbies were sometimes choked with them. To have borne that noisome atmosphere and heat would have tested the constitutions of salamanders, and engineers were only human.

Outside Parliament, he was constantly besieged by railway speculators. Samuel Smiles describes Robert Stephenson's offices in Great George Street as presenting 'very much the appearance of the levee of a minister of state'.

△ Colour lithograph by J C Bourne of a steam train crossing one of Brunel's earliest GWR bridges, at Maidenhead. Critics voiced doubts about the safety of its wide, elliptical arches, but it still carries heavy traffic today.

His father, George, hovered protectively by the door, always ready to engage in a war of words, or even a bit of arm-twisting.

In the early days of the railways, speculation had been confined to a few wealthy businessmen. But now ordinary working people, who could ill afford to spare the cash, were investing in shares, hoping for instant riches. Parliament was supposed to control development, but with the overriding ethos being that of the free market, lines were often granted that ran parallel to each other.

Support came from the very people who, just decades before, would have been most vocal in their opposition. Samuel Smiles noted:

One of the features of the mania was the rage for 'direct lines' which everywhere displayed itself. There were 'Direct Manchester', 'Direct Exeter',

*'Direct York', and, indeed, new direct lines between most of the large towns.
The Marquis of Bristol, speaking in favour of the 'Direct Norwich and
London' project, at a public meeting at Haverhill, said, 'If necessary, they
might make a tunnel beneath his very drawing-room, rather than be
defeated in their undertaking!' And the Rev. F Litchfield, at a meeting in
Banbury, on the subject of a line to that town, said 'He had laid down for
himself a limit to his approbation of railways, at least of such as
approached the neighbourhood with which he was connected, and that limit
was, that he did not wish them to approach any nearer to him than to* run
through his bedroom, with the bedposts for a station!'

In order to satisfy this voracious hunger for new railways, engineers had to
develop more efficient ways of working. Here, as so often, Brunel took the
lead. In the early days of the Great Western Railway, every bridge, viaduct
and station came directly from his drawing board. Perhaps the most dramatic
of these early constructions was the skew bridge at Maidenhead, which so
many dismal Jonahs were convinced would fall down.

As the railway progressed, Brunel continued to have a hand in features
he considered to be essential for the overall architectural effect – but he
also developed standard designs, which could be adapted as necessary. His
prototype country station, for example, can still be seen at Culham in
Oxfordshire, and variations on this survive throughout the GWR territory.

His vision encompassed every detail of the infrastructure needed for an
extensive railway network. This included the complex of workshops and
terraced housing for railway workers at Swindon. The site was chosen by
Gooch, as being on the junction with the Cheltenham and Gloucester line,
and near the point where the gradient became steeper. He noted in his diary
in 1840:

> *Mr Brunel and I went down to look at the ground, then only green fields,
> and he agreed with me as to its being the best place.*

Swindon was also the site of the infamous refreshment room – perhaps the
origin of jokes about the British Rail cup of coffee. At an early stage, Brunel
had made a rash agreement with a contractor that all trains should make a
10-minute refreshment stop at Swindon. Only first-class passengers were
admitted, but it does not sound as if this was a great honour. In December
1842, he wrote sarcastically to the manager:

▷ Another cartoon in
the Richard Doyle *Punch*
series, Manners and
Customs of Ye Englyshe,
showing that jokes
about railway buffets
are nothing new.

Dear Sir,

I assure you, Mr Player was wrong in supposing that I thought you purchased inferior coffee. I thought I said to him that I was surprised you should buy such bad roasted corn. I did not believe you had such a thing as coffee in the place; I am sure that I never tasted any. I have long ceased to make complaints at Swindon. I avoid taking anything there when I can help it.
Yours faithfully,
I K Brunel

This, no doubt, was the French side of Brunel's lineage coming out; it was also an example of his perfectionist nature.

Like Thomas Telford before him, Brunel was as much an architect as an engineer, and he revelled in every opportunity for the grand statement. However, he did not lose sight of practicalities. Much as Nash reserved expensive materials for the front of his Regency terraces in London, and used cheap stock bricks at the back, so Brunel put all his money into façades. So, for example, if you approach the Box Tunnel from the London end of the

GWR, you have an overwhelmingly grand vision of a classical arch, complete with balustrades that would not look out of place in the gardens of some country mansion. However, on the return journey from Bristol to London, the tunnel presents an extremely humble brick front.

Under Brunel, the biggest stations were transformed from functional train sheds into structures of pure fantasy. The original GWR terminus at Bristol Temple Meads, which now houses the British Empire Museum, had a dramatic, castellated front. Inside were iron columns supporting a Jacobean-style hammerbeam roof – a bizarre mixture of technologies if ever there was one. But it was at Paddington that Brunel found a form of architectural expression that truly suited the new iron age.

Both Brunel and Robert Stephenson were members of the committee that had been set up to oversee the construction of a suitable building to house the Great Exhibition of 1851. Brunel had, in fact, provided his own design for an enormous glass dome. However, he also admired the magnificent glass and iron structure by Joseph Paxton, the architect and landscape gardener at Chatsworth, which was the one that was eventually commissioned.

It was against this background that Brunel conceived his plan for a railway equivalent to Paxton's 'Crystal Palace' – Paddington Station. A temporary station already existed, but Brunel wanted something altogether more dramatic and more fitting for the greatest railway in Britain. He wrote to the architect Matthew Digby Wyatt with his plan for a series of inter-connecting iron-and-glass sheds, and asked somewhat gracelessly if he would be prepared to act as his assistant:

△ A bird's-eye view of Joseph Paxton's Crystal Palace, built to house the Great Exhibition of 1851.

> *I am going to design, in a great hurry, and I believe to build, a Station after my own fancy; that is, with engineering roofs etc. It is at Paddington, in a cutting, and admitting of no exterior, all interior and all roofed in. Now, such a thing will be entirely <u>metal</u> as to all the general forms, arrangements and design; it almost of necessity becomes an Engineering Work, but, to be honest, even if it were not, it is a branch of architecture of which I am fond, and, of course, believe myself fully competent for; but for detail of ornamentation I neither have the time or the knowledge, and with all my confidence in my own ability I have never any objection to advice and assistance even in the department which I keep to myself, namely the general design.*
>
> *Now in this building which* entre nous *will be one of the largest in its class I want to carry out strictly and fully all those correct notions of the use*

of metal which I believe you and I share (except that I should carry them still further than you) and I think it will be a nice opportunity. Are you willing to enter upon the work professionally in a subordinate capacity (I put it in the least attractive form at first) of my assistant for the ornamental details?

Digby Wyatt had worked on the Great Exhibition, and was one of the 'in' group of architects and designers who gathered around Prince Albert. He was a well-established professional in his own right, and many men in his position would have been offended at the terms of this offer. It says a lot for Brunel's reputation and personal charm that he accepted – especially when Brunel added: 'You are an industrious man and night work will suit me best.' A workaholic himself, he admired this quality in others.

Paddington Station has often been compared to a cathedral, with its massive central aisle and two side naves – and the recent renovation has brought out its soaring, architectural quality. It originally incorporated a canal basin, which was used both for transporting goods, and for bringing in

the vast quantities of coal needed each day. Every detail of its construction was carefully thought through. However, perhaps the most revolutionary aspect of the station was the close collaboration it represented between Brunel, Digby Wyatt, and the contractors, Fox and Henderson. This was the firm that had worked on the Crystal Palace, so they were well used to the demands of putting together buildings made of mass-produced components. Such collaboration is commonplace now, but was not then.

The iron and glass used at Paddington suits 21st-century tastes. However, it has not always attracted universal approval. Lady Noble, Brunel's grand-daughter, writing in the 1930s, pointed out the incongruity of such a grand structure for such a utilitarian function:

Perhaps it is just as well that, as we hurry through our particular station upon the business of arrival or departure, our more critical eyes should scarcely have time to notice the incongruity of engines puffing their smoke into a collegiate roof, of passengers, porters, and luggage jostling each other between the pillars of a nave, or of a balcony from which the directors may look down upon a train terminus instead of on the plashing fountains of an Arabic courtyard.

The station originally had no grand façade. However, this changed with the building of the Great Western Royal Hotel, designed by Philip Hardwick, which was completed in 1852. This was the first purpose-built railway hotel in Britain, constructed at a cost of almost £60,000. Its romantic French chateau-style details set the tone for Brunel's expansive ambitions. Next stop, not just Bath and Bristol – but New York.

Interestingly enough, Philip Hardwick was also commissioned to provide a gateway to Euston Station, the terminus to Robert Stephenson's London and Birmingham Railway. This was the massive Euston Arch – sadly demolished in the 1960s – and modelled on the great Greek and Roman monuments. Romantic and Classical; Brunel and Stephenson. Each now had a station to match their tastes.

△ Architecture meets engineering in Paddington Station, built by the same contractors as the Crystal Palace. Engraving from the *Illustrated London News*, 1854.

THE HUMAN COST

One of the Great Western Railway's best-known early passengers was the great romantic painter J M W Turner. Travelling up from Somerset in an almighty storm, he spent 10 minutes with his head out of the window, staring into the blinding rain as the train shot towards Paddington.

The resulting painting – *Rain, Steam and Speed* – is a swirling symphony of light, with the black funnel of the train the only recognizable feature in the picture. The painting was exhibited at the Royal Academy in 1844, and has since become an icon of the power of the steam age.

Mesmerized by the dramatic spectacle before him, it is unlikely that Turner spared much thought for the unnamed navvies whose sweat built the embankments and dug the tunnels through which he passed. The Victorians tended to ignore the men who got their hands dirty, and history has treated them little better. We still talk of achievements such as the building of the Great Western Railway as if they were the work of one person.

This was the age of the engineer as hero – and as showman. The sheer scale of these projects turned them into tourist attractions, inducing a mixture of admiration and terror. As science historian Simon Schaffer puts it:

'Engineering was theatre in this period. Vast numbers of visitors went deliberately to see a great Brunel bridge or a great Brunel boat or a tunnel. They were sublime objects and Isambard Brunel was a wonderful showman of the sublime.

'Sublime things are volcanoes, earthquakes, the Alps, a storm at sea, in which you feel two emotions, as it was said. You feel terrified because these natural forces are so much more powerful than mere humans are, but you feel very confident because you know we can master them.

'Brunel is a kind of P T Barnum of the sublime. He was very good at organizing spectacle so that you felt overawed by what he'd achieved, and yet full of admiration for what human beings like him could do.'

The banquets in the Thames Tunnel and the Great Western steamship, and the official opening of the Great Western Railway were carefully manipulated public relations events. However, both Brunel and Stephenson also had support from a whole publishing industry that rode on the back of the new technology.

In the early days of rail, artists painted pictures that happened to have trains in them; their aim was not to show the railway at work. That all changed with the opening of the Liverpool and Manchester Railway in 1830, when people began to hang prints of railway trains or the construction of the line itself in their drawing rooms. Some were meticulously detailed depictions of locomotives and their carriages; others emphasized the drama. These showed tiny figures toiling in the shade of massive rock faces, or trains chugging away like insignificant toys in a prairie-like landscape under a threatening sky.

The artist and engraver John Cooke Bourne (1814–96) continued in this tradition. The son of a London hatter, he first became interested in railways after observing the massive earthworks being thrown up by the London and Birmingham Railway less than a mile from his house. He started by making a series of sketches and watercolours of Stephenson's construction work, and went on to produce two books of lithographs, one on the London and Birmingham, another on the Great Western Railway. Both were in wide circulation by the early 1840s.

At a time when many people were still hostile to the disruption caused by railways, his work was a celebration of the audacity of the achievement. Like the early prints of the Liverpool and Manchester Railway, the exaggerated scale and stark contrasts of light and shade produce a feeling of Shelley's *Ozymandias* – 'Look on my works ye mighty and despair!'

However, in addition to these dramatic pictures, Bourne produced more modest sketches of everyday life. He shows horses pulling carts and eating

△ Previous page: Contemporary artists tended to exaggerate the scale of railway workings, for picturesque effect. This engraving by T T Bury shows the excavation of the Olive Mount Cutting on the Liverpool and Manchester Railway.

△ J M W Turner: *Rain, Steam and Speed – The Great Western Railway.* In this impressionistic portrayal of power and speed it is easy to miss the symbolism of the hare that dashes for cover in front of the advancing train.

from nosebags. He shows navvies putting up scaffolding, bricklaying, relaxing and sleeping. His interest is not only in the big set pieces of brick, stone and iron, but also in the teeming humanity behind them. It has been suggested that in another age, he might have become a photojournalist.

Engineering history tends to be dominated by the men who put their signatures on the plans. Brunel, in his diaries, writes about his *Chateaux d'Espagne* – castles in the air. They would have remained just that if it had not been for the army of individuals who not only made his dreams possible, but sometimes died for them. This was true of Brunel, as it was for every other railway engineer in the country.

Behind the cuttings, tunnels and embankments were assistant engineers and contractors, and navvies with iron muscles and nicknames like Hedgehog, Gorger, Concertina Cockney and Rainbow Peg. Rainbow Peg got his name because he had lost one leg, and put all his weight on the other, which was consequently bowed like a rainbow. Peg, Peggy or Wingy were common nicknames for those who had lost an arm or a leg.

Maiming or mutilation came with the job, and navvies were lucky if they escaped with nothing more than the loss of a limb. They worked using picks and shovels, crowbars and wheelbarrows, and their bare hands; the only other aid they had was the occasional blast of gunpowder. Some were blinded by explosions; others were buried in rock falls. All led a life of hard, grinding physical toil, tramping from one construction site to another in search of work. Their reputation for violence and drunkenness made them a frequent focus for missionaries and temperance society members, as well as turning them into the bogeymen of folk myth. 'Whist, or I'll give yer to a navvy,' was a threat used by Yorkshire mothers to their children.

While some people admired their brute strength, the word 'navvy' was for many an insult. Peter Lecount, Assistant Engineer on the London and Birmingham Railway, described them as having:

NAVVY IN HEAVY MARCHING ORDER.

△ *Punch* cartoon (1855) showing the popular stereotype of the army of navvies who tramped from one railway construction site to another.

> … all the daring recklessness of the Smuggler, without any of his redeeming qualities. Their ferocious behaviour can only be equalled by the brutality of their language. It may truly be said, their hand is against every man, and before they have been long located, every man's hand is against them.

In a contemporary account of the history of railway construction, *Our Iron Roads* (1852), one Frederick S Williams noted that:

> … a frightful picture must be sketched of the condition of large numbers of the navvies, during the period of the construction of so many lines. Living together in some places like herds of brutes – working on the sacred hours of the Sabbath – destitute of instruction, either for themselves or their wretched offspring – subject at every hour of the day to frightful accidents – disregarded by others, and careless for themselves; the last relic of civilization seemed to disappear, as they even changed their names into the

uncouth and barbarous epithets by which they preferred to be known.
Painful is it to find that the triumphs which the human intellect has
achieved should be so intimately associated with the moral degradation of so
large a section of the community.

'Navvy' was an abbreviation of 'navigators'. It was first applied at least as early as 1775 to the men who worked on the canals, or 'inland navigations'. There were no instruction manuals explaining how to form cuttings, embankments and tunnels – people just went to the nearest construction site to see how it was done. As railways took over from canals, the navvies transferred their skills. In every year of the 19th century, from the mid-1820s onwards, railway work was available for navvies somewhere in Britain. At the height of the construction boom, in 1847, over 200,000 men were employed on manual labour in railways throughout Great Britain. This was greater than the strength of the armed forces, which at that time stood at 160,000 men.

The popular stereotype of the navvy is of an Irish labourer, living out a feckless, drunken existence in a series of temporary lodgings. In fact, many of the labourers who worked on the railways were local men, attracted by the prospect of wages that were higher than could be offered on neighbouring farms. The 1841 census noted that of the 1,255 people on the Wootton Bassett to Box section of the Great Western Railway, 26.9 per cent had been born in Wiltshire, and only 1.9 per cent in Ireland. However, in Scotland, which saw more immigration from Ireland than either England or Wales, the story was different. For example, the Irish were the leading railway builders in the borders area.

Shanty towns sprang up around the more remote rural railway workings, where it would have been difficult to recruit local labour. These ranged from simple huts made from turf to much grander affairs. The London and Birmingham Railway put up cottages and sheds at Blisworth, Kilsby, Tring and Harrow in 1836 and 1837. The contractors had been so desperate for extra hands at Blisworth Cutting, in Northamptonshire, that they posted scouts on all the nearby roads, with instructions to offer any male pedestrian a job. The cottages at Blisworth boasted fireplaces and outside privies – and the specifications even prohibited the use of wood with large knots for floorboards. The weekly rent was 4s 6d per man (22½p), with free supplies of coal.

'I think navvy wives were more remarkable than navvy men,' says railway historian Adrian Vaughan. 'Where the navvies were congregated in huts or other encampments the women bore children, nursed them, defended them when drunken brawls broke out, tried to keep order with fists or broom handles and had their limbs broken by the irate males. The men's wives kept the huts clean and cooked for several men and were paid a little by each. Or maybe some women succumbed to despair and drunkenness.'

The missionary Anna Tregelles described an encounter with Sarah, the wife of a Box Tunnel navvy known as 'Chimley Charlie'. She wore a black bonnet with bright ribbons, earrings, a light muslin dress, and stout leather boots. She could not read or write, as her family had never been able to find the weekly twopence for the village dame's school. However, she knew her letters – 'higley-pigley, not straight on' – and was clearly well able to look after herself.

Sarah said she was thankful that she had never had any children, 'for they were the ill-convanientest things that ever was to a navvy woman'. She did not explain how she had managed to avoid conceiving – but it was obvious that she had the upper hand in the marriage. As she told Miss Tregelles:

> *There, Miss, that's the mark where he laid my hand all open with his strap; but I up and knocked 'em down with the poker. He come down sich a bang, that I thoft I'd a killed 'en, and I screeched like mad. Yes, us have had fights since, but he've never struck me with nothing but his hand again. 'Twas all about the money as he drank.*

Because they were often in isolated encampments, and because they sometimes conversed in a strange rhyming slang all of their own, navvies had the reputation for being social outcasts. They were frequently made scapegoats for local disorder in which they were only marginally involved. In June 1847, the directors of the South Devon Railway received a bill for £46 to cover the cost of enrolling over 300 special constables during two days of disturbances on Torquay the previous month. However, it transpired that the trouble had begun as a bread riot in which the railway labourers were a very small group. Some of the 500-strong crowd had imitated navvy language, and the navigators had therefore been blamed.

Brunel himself was guilty of inciting navvies to violence in the so-called Battle of Mickleton Tunnel in 1851. He had fallen out with the contractor in charge of the tunnel, on the Oxford to Worcester line. The contractor and his workmen barricaded themselves into the half-finished workings, and Brunel raised a gang of railway navvies to evict them. After a stand-off,

△ These realistic sketches by J C Bourne celebrate the achievements of navvies (and horses) on the railways. Note the figure of the overseer, hands in his pockets, as the navvies toil with their pickaxes.

during which a local magistrate read the Riot Act, Brunel retreated. However, he later returned with a force of some 2,000 navvies against a band of only 150 workmen. The episode was eventually resolved peaceably (in fact, Robert Stephenson was one of the arbitrators) but Brunel narrowly risked losing his reputation through these bully-boy tactics.

The life of a navvy was hard, brutal, and often short. The first recorded death of a railway navvy was in the spring of 1824, when one W Mossom, a labourer on the Stockton and Darlington line, was killed. *The Liverpool*

Mercury for 10 August 1827 noted a fatality on the Liverpool and Manchester Railway:

> *We are pained to state that a labourer, who was working in the excavation for the rail-road, at Edgehill, where the tunnel is intended to come out and join the surface of the ground, was killed on Monday last. The poor fellow was in the act of undermining a heavy head of clay, fourteen or fifteen feet [4.3-4.6 m] high, when the mass fell upon him, and literally crushed his bowels out of his body.*

Accidents often happened when contractors or navvies tried to speed things up. Contractors generally worked to deadlines, and if – as sometimes happened – the navvies were on piecework, both groups had a vested interest in going faster. One extremely risky technique was sometimes used when making a cutting. One group of men would drive in piles at the top of the excavation while, at the same time, another group would hack away at the bottom. Success depended on knowing when to stop; this in turn depended on experience and local knowledge of the texture of the soil. If it worked, the

△ *Work*, by Ford Madox Brown, is unusual for its time in placing workers at the heart of a painting. The artist used genuine navvies as his models, and one of them was killed in a scaffolding accident before the picture was finished.

whole middle section would fall away in one piece; if it failed, those working at the bottom would be crushed to death. The contractor John Sharp, giving evidence to the 1846 Select Committee on Railway Labourers, recalled one incident in which a new and promising worker was repeatedly told not to continue to 'knock the legs' from under a certain section of soil. Minutes later, the rock face collapsed and buried him.

Old hands played tricks on new recruits, such as overloading their wheelbarrows. On a 'barrow run' such as that at Robert Stephenson's 40-ft (12-m) deep chalk cutting at Tring, only the most experienced men had the balance and co-ordination necessary to wheel a loaded barrow up and down the slippery planks. A rope, attached to the barrow, and also to each man's belt, ran up the side of the cutting and then round a pulley at the top, where it was attached to a horse. If, on the upward climb, either the horse or the man slipped or faltered, there was a real danger that the barrow would overturn, trapping the man underneath it. Navvies got used to being thrown down the slope, and, miraculously, only one was killed in this way. In fact, navvies seemed to accept the risk as an integral part of the job. When a moving platform was invented to make their work safer, they saw it as a threat to their livelihoods, and smashed it.

Macho recklessness was a navvy trademark. When the Kilsby Tunnel was under construction, two or three workmen were killed trying to jump one after another in a game of 'Follow my Leader' across the mouth of the shafts. Further along the London and Birmingham line, at Blisworth Cutting, the workmen developed the habit of hitching a lift with the tip wagons, which took the spoil along to a nearby embankment. These wagons frequently ran off the rails, pitching the workmen and stone alike on to the ground. In *Our Iron Roads*, Frederick Williams described the men's reactions:

On one of these occasions, though but a few days after a fatal accident of a similar nature had taken place, some wagons were thrown off the rails, and several men buried beneath the limestone. One stalwart fellow scrambled out from the heap, and feeling his arm, said to a more fortunate comrade – 'It's broke, I maun go home;' and, having waited only to ascertain the fate of his fellow-sufferers, he strode off to his dwelling, which was six miles distant, supporting the broken limb with the sound one.

A fine handsome youth, who by the same accident had his foot crushed into a shapeless mass of flesh and bone, gave vent to his feelings by crying bitterly. A rough-looking ganger who stood by, took the pipe from his lips, and in a blunt, advising way, said to the boy – 'Crying 'ill do thee no good, lad;' and then, as if somewhat acquainted with the mysteries of the scalpel, added, 'thou'dst better have it cut off above the knee.'

The large number of casualties from railway workings was a burden on hospitals. Contractors and railway companies were supposed to contribute towards the cost, but they rarely paid enough. The London and Birmingham, in addition to supplying a handcart for carrying the injured, contributed only £115.10s to Northampton General Infirmary. The hospital had, in fact, treated 124 patients, at a cost of nearer £600. The story goes that Brunel was once shown a list of 131 navvies who had been taken to Bath Hospital from 30 September 1839 to 24 June 1841. His only comment was that the list was smaller than he would have expected.

'Brunel never bothered about navvies,' says Adrian Vaughan. 'It was a ridiculous failing because the navvies were building his railway and therefore he ought to have had more concern, because his glory depended on them doing the job. But Brunel said, they don't have to work for me, I don't ask them to, and so I'm not too bothered about them.'

Railway directors kept a careful tally of accidents involving passengers, but when it came to labourers, it seems they were a little more cavalier. Many accounts put the death toll during the building of the Box Tunnel as high as one hundred. However, it is impossible at this distance to know exactly how many men were killed. David Brooke, the author of *The Railway Navvy, 'That Despicable Race of Men'*, has traced this figure to a pamphlet written by the superintendent of the tunnel works. The pamphlet notes that 'just one hundred men were killed in different parts of the works and in different ways', which could refer to casualties on other parts of the line as well as the tunnel. Research in local archives has turned up evidence of only (!) nineteen deaths, with two more not expected to survive. Most of these casualties were workmen who had fallen down tunnel shafts, although one 21-year-old was killed instantly when a skip load of bricks fell on him from 200 ft (61 m).

Conditions in a 3-mile (5-km) long railway tunnel being excavated under the Pennines for the Manchester and Sheffield Railway were so bad that they led to the establishment of the Parliamentary Select Committee on Railway Labourers. The notorious Woodhead Tunnel was built over five years by a force of over a thousand navvies. The men lived with their wives and children in heather-roofed huts on bleak moorland, working long hours often knee-deep in mud and water. Many had chronic coughs, which they blamed on the

△ Illustration by J C Bourne showing the barrow run at Tring, on the London and Birmingham Railway. One slip could be fatal.

moistness of the tunnel; their clothes were wet through within a quarter of an hour of starting work, and never properly dried out.

The committee was told of one man who lay for several days with a broken back; he continually asked for a clergyman, but no one would agree whose parish he was in. By the time a Methodist minister agreed to visit him, he had died. One of the two surgeons who attended the injured men listed thirty-two deaths, as well as twenty-three cases of compound fracture, including two fractured skulls, seventy-four simple fractures and one hundred and forty other serious cases, including blast burns, contusions, lacerations and dislocations. One worker lost both eyes and another man half his foot. Some four hundred other accidents included trapped or broken fingers, seven of which had to be amputated.

When the tunnel was finished, the cost in money was £200,000 – almost exactly the original estimate put forward by Joseph Locke, the engineer. However, as the social reformer Edwin Chadwick pointed out, the cost in human lives in proportion to the total force engaged had been greater than that of any battle of the century, Waterloo not excepted. As L T C Rolt put it: 'It displays that age at its best in the sheer pertinacity with which a task of incredible difficulty was tackled, and at its worst in the appalling conditions and in the contempt for human life shown by contractors and resident engineers alike.'

There is no doubt that tunnelling was easily the most hazardous occupation undertaken by the navvies. Working underground for as much as twelve hours at a time, the men were in constant danger from the explosions set off to blast through the rock, from rock falls and from flooding. It must also have been a noxious atmosphere to work in, with fumes from the gunpowder mixed with the stale underground air.

In a longer tunnel, shafts would be bored with the help of a machine called a gin. This took its name from the middle part of the word 'engine', and was powered at first by horses attached to a great wheel, and later by steam engines. Miners would go down sometimes as deep as 600 ft (183 m) into the ground, and the same buckets would be used to send up spoil from their workings.

At the bottom of the shaft, the men would tunnel in two directions, using wooden props to support the roof until it could be lined with bricks. Danger sometimes came from unexpected quarters. Ten men were crushed to death in the Watford Tunnel on the London and Birmingham Railway after a shaft fell in, taking with it the gin, the man who had been working it and his dog. It turned out that the shaft had been sunk next to a fault in the chalk. When the men underground removed the wooden supports, the tunnel collapsed around them.

Brunel's Box Tunnel, the longest railway tunnel of its time, had attracted much attention from armchair critics at the planning stage. Dr Dionysius Lardner raised fears about train noise, and said that no passenger would be induced to pass through it twice. Another of Brunel's opponents suggested the tunnel was 'monstrous, extraordinary and impracticable' and would cause the 'wholesale destruction of human life'. People attacked both the length and the gradient (1 in 100), and there were questions about the suitability of the rock that it passed through.

It is hard to work out exactly what caused the furore. Robert Stephenson's Kilsby Tunnel went through equally difficult terrain, although admittedly it was 786 yards (718 m) shorter than the tunnel at Box. True, there was the question of the slope – but there is a feeling perhaps that people thought Brunel was being a sight too clever for his own (and other people's) good.

As usual, Brunel found it difficult to delegate. However, even he realized it would be impractical to supervise every inch of the tunnel himself. Writing

△ A horse is dramatically illuminated by a ray of sunlight in J C Bourne's depiction of a working shaft on the Kilsby Tunnel, 8 July 1837.

in March 1836 to William Glennie, who had applied for the post of Resident Engineer, he made it clear who was boss:

My responsibility is too great to allow of my retaining for one moment from any feeling of personal regard, the services of any one who may appear to me to be inefficient from any cause whatsoever and consequently it is an understood thing that all under me are subject to immediate dismissal at my pleasure. You will perceive that I state all these conditions in strong and perhaps harsh language and that in substance they are exactly what I stated to you.

Brunel's early experience of tunnelling – under the Thames – had taught him to be wary of geological surprises. He was therefore careful to sink trial shafts before parcelling out the work to contractors, an early example of a site investigation that was ahead of its time. (Robert Stephenson did the same thing before building the Kilsby Tunnel, although his shafts did not reveal the quicksand that was to bedevil his work there.) Brunel also paid a visit to the Whitstable to Canterbury Railway, where a tunnel with a 1-in-100 gradient had been successfully worked by trains.

With a downward slope running towards Bath, the tunnel passed through layers of oolite (a Jurassic limestone famous as Bath stone), Fuller's Earth and clay. Brunel would have been able to visit quarries of Bath stone, so would have seen for himself its durable qualities. He used it to line the tunnel's grand eastern entrance, and left bare rock for the roof and sides of those sections that passed through oolite. Seven shafts, 25 ft (7.6 m) in diameter, were sunk; as always, Brunel worked to generous measurements.

The tunnel was the last link in the GWR line from London to Bristol, and work was slow in starting. Adrian Vaughan says part of the reason for this was that Brunel was notoriously parsimonious, and found it difficult to find a contractor who would work with him. The experienced contractor George Burge, of Herne Bay, took on the contract for three-quarters of the length of the tunnel. He had previously worked on St Katharine's Dock, London for Thomas Telford. His section was to be lined with brick throughout, and as some thirty million bricks would be needed in total they were made nearby, at a yard in meadows to the west of Chippenham. The unlined section at the eastern end was undertaken by two local men, Brewer of Box and Lewis of Bath, who were used to working with Bath stone.

Once the shafts had been sunk, the first stage was to drive a pilot heading, and then enlarge that cut by working downwards. There was room for only three or four men in the heading, and conditions were cramped and dangerous. Gunpowder was used to blast through the rock – a whole ton of the stuff was used every week. The riskiest job was tamping the powder. The men would drill a hole in the rock, fill it with gunpowder and then hammer a clay plug in to compact the powder. When this process was reconstructed for the *Men of Iron* TV series, explosives expert Sidney Alford used a wooden tamping rod (or 'stemmer', as the Box Tunnel navvies would have called it). However, this would originally have been iron. The danger was that the smallest scratch of the rod on rock might cause sparks; this would set off the gunpowder and fire the stemmer straight through the head of the unfortunate navvy.

The holes for blasting would have been made by hand-held drills. The hammer man would strike while the holder-up would rotate the drill after each stroke. Relays of small boys, called tool carriers – usually sons of miners – would run to and fro to take the drills to the blacksmith for constant re-sharpening.

'They would have spent hours hammering to make a hole big enough to pack with gunpowder,' explains Adrian Vaughan. 'Then when they explode the charge they go in to shovel up the rock and the place is full of gunpowder fumes. It's going to be stinking, and until the tunnel is complete enough to be open at the ends, it's like being in a sealed-off cave.

'So these men are working in candlelight, in gunpowder fumes, in freezing cold, water running in as it comes through the soil – very difficult indeed.'

Water was a constant problem, especially at the end worked by Lewis and Brewer. In November 1837 one of the shafts was flooded to a height of 56 ft (17 m), and work had to stop for some months until a second steam pump could be installed. However, apart from steam pumps such as these, all the work was done by human muscle and horsepower. No doubt economics dictated the use of horses rather than steam engines. It would have been cheaper to harness a couple of horses to tread in endless circles round a gin rather than use indefinite numbers to haul coal across inadequate country roads to power steam engines.

The excavated spoil was hauled up to the surface in open buckets, and well-dressed visitors were taken down into the bowels of the tunnel in those same muddy buckets, or on open platforms. Just as the Thames Tunnel had proved a popular tourist attraction, so people were curious to see this strange subterranean world. In 1838, a reporter from the *Wilts Independent* descended 136 ft (42 m) into 'shaft no. 7' and described the dramatic moment of a gunpowder explosion:

> *The match is applied, the explosion follows, and a concussion such as probably you never felt before takes place, the solid rock appears to shake and the reverberation of the sound and shock is sensibly and fearfully experienced; another and another to follow, and with a slight stretch of the imagination you might fancy yourself in the midst of a thunder cloud with heaven's artillery booming around.*

The Bath-stone end of the tunnel was worked by local quarrymen, but the rest of the excavation needed a much larger workforce recruited from outside the area. From early 1838 onwards, gangs of navvies began appearing in the district, many from finished sections of the London end of the Great Western line. Some lived in makeshift cabins near the tunnel; in summer, some simply slept in the open in cuttings. However, most lived as lodgers in the nearby villages. They worked shifts, and no bed ever went cold.

On Sundays, the foremen had to go out into the villages to keep the peace. This must have been an immense invasion into the quiet countryside.

At its peak, the tunnel occupied a force of some 4,000 men and 300 horses; they used a ton of candles a week. Sections of it were extremely muddy, making the work back-breakingly slow and laborious. The navvies' shovels would become dead weights as they tried to offload the sticky mess into trucks. Chimley Charlie, Box Tunnel's best-known navvy, finally gave up in despair, having worn through the soles of two pairs of boots in the morass. He went off to Leeds, in search of some rock to shovel instead.

△ Despite the grand classical entrance at the West end of the Box Tunnel, some passengers were too nervous to penetrate its dark depths. Lithograph of 1846 by J C Bourne.

Because tunnelling was carried out in several sections at once, there was always the fear that a gang might veer off course. The Bath contractors Brewer and Lewis started working at opposite ends of their section. Brunel was present at the moment when the two ends met, and the story goes that he took a ring from his finger and gave it to the foreman in token of his gratitude.

The tunnel was finally completed in June 1841, and now the whole route from London to Bristol was open. That same year, excavators on the Thames Tunnel finally broke through into the Wapping shaft, and Marc Brunel's grandson, the three-year-old Isambard, was handed through a driftway to become the first person in history to pass beneath the Thames. (This was the Isambard who was later to become his father's biographer.)

It might be thought that the successful completion of both tunnels would silence the critics. However, this was not so. There were still people who refused to go through the grand classical portals of the Box Tunnel into its dark, sulphurous interior. They would get off at the entrance, take a coach along the turnpike road, and wait for another train at the other end. It had originally been planned to light the tunnel with reflector lamps. In the event, this was found to be impracticable, as the atmosphere was too smoky for the lights to be of any use.

At the opening, only a single line through the tunnel was ready. For the first 48 hours Daniel Gooch, the Locomotive Superintendent, travelled as a pilot with every train. In his diary, he recalled some anxious moments:

At about 11 o'clock on the second night we had a very narrow escape from a fearful accident. I was going up the tunnel with the last up train when I fancied I saw some green lights in front. [All GWR trains carried a green headlight.] A second's reflection convinced me it was the Mail coming down. I lost no time in reversing the engine I was on and running back to Box Station with my train as quickly as I could, when the Mail came down behind me. The policeman at the top of the Tunnel had made some blunder and sent the Mail on when it arrived there. Had the Tunnel not been pretty clear of steam, we must have met in full career and the smash would have been fearful, cutting short my career also. But, as though mishaps never come alone, when I was taking my train up again, from some cause or other the engine got off the rails in the Tunnel, and I was detained there all night before I got all straight again. I need not say I was not sorry to get home and to bed at Paddington, after two days and nights pretty hard work.

Rumours about the safety of the tunnel were reinforced a year later by alarmist comments from Dr William Buckland, an Oxford geologist, at a meeting of the Institution of Civil Engineers. Without ever even having visited it, he claimed that the vibration of passing trains might loosen large pieces of oolite, which could fall and cause a serious accident. Brunel could not afford to ignore the opinion of such a learned man, however misguided he thought him. His carefully worded letter conveyed a plain message, yet ended with a calculated piece of polite deference:

> *I feel that as regards the works of the Box Tunnel everything necessary has been done to render them secure and that the doubts and fears you have so easily raised but which it might be more difficult again to set at rest, are entirely unfounded. In conclusion I must observe that no man can be more sensible than*

△ By the end of the 19th century, human earthmovers were beginning to be replaced by machines such as the 'Steam Navvy'. This photograph, commissioned by one of Brunel's successors as Chief Engineer to the Great Western Railway, was taken in Wiltshire in 1899.

I am of the great advantage it would be to me as a civil engineer to be better acquainted with geology as well as with many other branches of science.

In the short term, his best argument was the very fact that trains were passing through the unlined section of the Box Tunnel without the predicted avalanche. However, in March 1845, a severe frost brought down a piece of stone weighing about three hundredweight (150 kg). The stone derailed a light engine, although nobody was injured. As a precaution, part of the tunnel was lined with brick.

Meanwhile, what of the navvies who had worked so hard to join the last link in Brunel's 'billiard table'? As was their way, they moved on to other jobs. After the establishment of the 1846 Select Committee, many parliamentary hours were taken up with analysis of the grimness of their working conditions. Engineers, including Brunel, gave evidence, as did contractors, policemen and railway missionaries. Only three navvies were invited to speak, and they received a mere 171 questions out of more than 3,000 put to other witnesses.

Most agreed on the evils affecting navvies. Jobs were too often sub-contracted to fly-by-night employers, who paid their workers irregularly, or not at all. Sometimes they were paid not in money, but in food and drink. The notorious 'truck system' meant part of their pay was in tickets to be exchanged at over-priced 'tommy shops' run by contractors. ('Tommy' was navvy slang for food.) If they were injured, neither they nor their families would get compensation unless the contractor happened to be particularly benevolent, or there was a sick-pay fund.

Reformers hoped for some form of parliamentary legislation to protect navvies. In the event, nothing happened. The committee produced a report that got lost amid the debate over the repeal of the Corn Laws and was never even discussed. As railway mania subsided, fewer railways were built and there was a diminishing need for human earthmovers. By the 1890s, it looked as if the mechanical excavator – or 'steam navvy' – was set to take over. However, in the First World War, it was navvies who dug the trenches for their mates to die in. Even in the supposedly enlightened 20th century, Chimley Charlie's nameless successors still had a role to play.

BACK FROM THE BRINK

'**M**y courage at times almost fails me; and I fear that some fine morning my reputation may break under me like an egg-shell.'

So the 26-year-old Robert Stephenson confided to an old friend from his days in South American; he was not to know how prophetic these words were. A quarter of a century later, a train bound for Shrewsbury steamed out of Chester Station and made to cross Stephenson's bridge over the River Dee. It had almost reached the far bank when, with a thunderous crack, a girder gave way beneath it. The carriages, complete with passengers, plunged into the river 36 ft (11 m) below.

The engine driver had a miraculous escape after his locomotive became detached from the rest of the train. He had put on extra steam after feeling an unusual vibration, and had just reached firm ground when the bridge broke. His fireman, who was breaking coal on the tender at the time, was not so lucky. Neither were the guard or the two coachmen who were travelling in the leading van. All four were killed instantly as the train hurtled down on to the riverbed. A fifth man died later of his injuries, and sixteen others were hurt.

At the inquest that followed, Robert Stephenson was lucky to escape a verdict of manslaughter. At the inevitable Royal Commission, which was set up after the disaster, he was lucky to escape with his reputation. Only the support of his fellow engineers, Brunel among them, ensured that his career would continue on its upward path.

The successes of engineering history are all around us, and this makes it easy to forget the failures. Perhaps because for so many Victorians life was nasty, brutish and short, they accepted a far higher degree of risk than we would even contemplate today. It could be argued that it was only because they had what we would see as a fairly callous attitude to individual human life that they were able to make such great strides in engineering.

Super-cautious engineers like Thomas Telford, with his home-made testing apparatus in the meadows next to the Menai Straits, experimented with materials as far as was possible. Robert Stephenson – like Telford, and unlike Brunel – tended to err on the side of prudence. However, in the end, the only way of being sure a particular bridge would stand up to the stresses and strains of daily traffic was to build it and try it out in practice. Each new construction was an experiment, testing the capabilities of different materials to their limit.

The irony of the Dee Bridge disaster was that it may have been precipitated by a measure taken to make it safer. The bridge was made of cast iron, with wooden planking underneath the rails. Brunel had used cast iron and wood for a bridge over the Great Western Railway at Hanwell that had caught fire. The heat of the fire had destroyed the cast-iron girders. No one was injured, but the episode so alarmed officials of the Chester and Holyhead Railway that they laid eighteen tons of broken stone to protect the wooden surface of the Dee Bridge. This was on the afternoon of 24 May 1847. The first train to go across the tracks afterwards was also the last.

Whether or not the extra ballast laid on the tracks had indirectly caused the bridge to fail, it soon became clear that there had been an underlying weakness. After the accident, a broken girder was retrieved from the riverbed, fractured into three pieces.

Robert Stephenson had built other similar bridges, with cast-iron beams resting on stone piers. In fact, he had built more iron bridges than any other railway engineer, so he was not lacking in experience. The only major difference between this and his previous constructions was that he used longer girders; it was a three-span bridge, with groups of four girders to each span, each one 98 ft (30 m) long. However, the bridge had been inspected and passed for traffic by Major General Sir Charles Pasley, the Board of Trade Inspector General of Railways, in October 1846. Trains had been running across it without incident for seven months.

△ Previous page: *Conference of Engineers, Britannia Bridge*, a painting by John Lucas. Robert Stephenson is seated at the centre of the table, with his hand on the plans. Joseph Locke is on the chair with the dog underneath, and Brunel is on the extreme right.

In his memoir, the engineer F R Conder is highly critical of the inquest that was held to establish the cause of the accident. As he put it:

The most clear and indubitable result of this investigation was, to demonstrate the entire unfitness of the ancient machinery of the coroner's court, presided over by a country doctor or lawyer, in which solicitors appeal to the convictions of an incompetent jury, unchecked by the presence of a bar, to deal with questions involving such important issues.

The foreman of the jury was Sir Edward Walker, sometime Mayor of Chester. According to Conder, the inquest at times almost degenerated into a personal contest between Walker and Robert Stephenson, who was understandably rattled by the proceedings:

Pale and haggard, he [Robert Stephenson] looked more like a culprit than like a man of science, assisting in a painful investigation. His manner was abrupt and dictatorial, betraying extreme irritation at the remarks of the jurors; and on more than one occasion he attempted, on the score of professional knowledge, to put down with some contempt the questions and remarks of Sir E Walker, who, nettled in his turn, affected to treat Mr Stephenson as a culprit on his defence.

It is clear from Robert Stephenson's comments later in life that he understood perfectly the real reason for the accident – that cast iron was simply the wrong material for long horizontal girders. However, he was forced to go along with the elaborate scenario constructed by the Chester and Holyhead Railway Company's lawyer. This was that a wheel had broken on the tender, knocking it off its tracks, and this had jolted the bridge and broken its back. In other words, the blame was put on the train, not the bridge. Stephenson and Joseph Locke, an equally eminent railway engineer, both gave evidence in support of this theory.

Locke showed considerable loyalty to a fellow engineer in giving this evidence, as he himself avoided using cast iron in his bridges; preferring to use brick or stone wherever possible. Cast iron, as the name suggests, is iron that has been poured or cast into a mould. It is strong, but brittle; if you hang too heavy a weight off a cast-iron bar, it will break before it bends. The longer the bar, the weaker it gets, and in the Dee Bridge, Stephenson had taken it beyond its limit.

Stephenson knew cast iron had its limitations as a material. He therefore reinforced his cast-iron beams with rods of wrought iron, a tough, malleable form of iron suitable for forging or rolling. The principle was the same as that used today in reinforced concrete. However, unfortunately for him, the method of attaching the rods to the beam was unsound, and the reinforcement did not work. In fact, one engineer who gave evidence at the inquest suggested that the wrought-iron truss rods actually weakened the girders by adding extra local stresses.

Major General Pasley, the Board of Trade Inspector who had passed the bridge, was so thrown by the whole procedure that his evidence was almost inaudible. However, he pointed out that a similar bridge over the Tees at Stockton, which used girders with a span of 87 ft (26.5 m), had been working successfully; there was therefore no reason to suppose the Dee Bridge would be any different.

In the event, the jury did not find anyone guilty of manslaughter, but they rejected the company's defence. They concluded that the girder did not fail

△ The Dee Bridge Disaster. Engraving from the *Illustrated London News*, 1847.

due to any blow, 'but from its being made of a strength insufficient to bear the pressure of quick trains passing over it'.

They called for a parliamentary inquiry into the use of iron in railway bridges, so a Royal Commission was duly set up, 'for the purpose of inquiring into the conditions to be observed by engineers in the application of iron in structures exposed to violent concussions and vibration'. Brunel was strongly opposed to any interference in engineering activities from on high. Like Joseph Locke, he disliked cast iron, and rarely used it. If he did, he was very particular about the mixture used, and made sure that an engineer supervised the casting. After the bridge at Hanwell burnt down, he had it repaired with wrought iron. However, he refused to condemn Robert Stephenson's use of cast iron, and dubbed the inquiry 'the Commission for stopping further improvements in bridge building'.

Having tried in vain to be excused from giving evidence, on the grounds that it would be 'productive of much mischief, both to science and to the profession', a large part of his written evidence was a plea against such inquiries:

> *Nothing, I believe, has tended more to distinguish advantageously the profession of engineering in England and America, nothing has conduced more to the great advance made in our profession and to our pre-eminence in the real practical application of the science, than the absence of all regles de l'art – a term which I fear is now going to be translated into English by the words 'conditions to be observed'. No man, however bold or however high he may stand in his profession, can resist the benumbing effect of rules laid down by authority. Occupied as leading men are, they could not afford the time, or trouble, or responsibility of constantly fighting against them – they would be compelled to abandon all idea of improving upon them; while incompetent men might commit the grossest blunder provided they followed the rules. For, in the simplest branch of construction, rules may be followed literally without any security as to the result.*

Brunel deliberately avoided the issue of Stephenson's bridge, choosing instead to generalize:

> *At present cast iron is looked upon, to a certain extent, as a friable, treacherous, and uncertain material; castings of a limited size only can be safely depended upon; wrought iron is considered comparatively trustworthy, and by riveting, or welding, there is no limit to the size of the parts to be used.*

Yet, who will venture to say, if the direction of improvement is left free, that means may not be found of ensuring sound castings of almost any form, and of twenty or thirty tons weight, and of a perfectly homogeneous mixture of the best metal?

Officially, Robert Stephenson's name was cleared, but he knew that he had made an horrific blunder and narrowly escaped professional ruin. Things were made worse by the fact that he now faced the biggest challenge of his professional career. Just months before the Dee Bridge collapse, he had been making plans for the most ambitious bridge he – or anyone else – had ever built. As fate would have it, the new construction was to be on the same line as the Dee Bridge – the Chester and Holyhead Railway. He admitted afterwards: 'I stood on the verge of a responsibility from which, I confess, I had nearly shrunk.'

The new bridge was to cross the treacherous Menai Straits – a site that already had its place in engineering history. In fact, when the railway was first discussed, there was talk of using one of the two roadways of Telford's great suspension bridge. This was abandoned after doubts about whether the bridge could stand the vibrations and weight of locomotives.

Stephenson's original plan had been for a bridge of two cast-iron arches, using the Britannia Rock – a small island in the middle of the stream – as a stepping stone. However, even before the Dee Bridge disaster, he had abandoned the idea of using cast iron. Also, the Admiralty insisted that any bridge should have a clear passage of at least 100 ft (30 m) to allow tall ships to pass underneath. Stephenson toyed with the idea of making a specially strengthened suspension bridge. But the same problem arose: what material should be used to build it?

Wrought iron was undoubtedly the best available bridge-building material. It did not carry the same risk of fire as timber, or the brittleness of cast iron. To make it, you had to melt the raw material in a furnace and then hammer it while it was still hot. Bashing the iron forced out its impurities, while folding it repeatedly under the hammer gave it a multi-layered consistency. All this made wrought iron stronger, yet more elastic than cast iron. But because there was a limit to how big the furnace could be, there was also a limit to how big a piece of wrought iron could be produced.

The solution to Stephenson's dilemma came from an unexpected source – a shipyard accident. A brand new steamship called the *Prince of Wales* was being launched at Blackwall when, halfway through the procedure, the launching gear broke. The ship's bow got stuck on a wharf, and in the process of pushing it off with jacks, the hull was left hanging unsupported in mid-air. Any other ship would have broken her back, but the *Prince of Wales* was not any other ship. She was made of iron. Her skin of wrought-iron plates was only half an inch thick, but the plates had been riveted together to form a stiff hollow tube – the hull – which turned out to be surprisingly strong.

Encouraged by this, Stephenson came up with the idea of constructing a great hollow tube out of wrought-iron plates, and making it big enough for a train to pass through. The tube would provide support, as well as a passage for the railway, all at the same time. But – a question close to his heart after the Dee Bridge disaster – would it be strong enough?

To find the answer, he turned to two people: the scientist Professor Eaton Hodgkinson and the shipbuilder William Fairbairn. This was early days for wrought iron, and shipbuilders were the only people with enough experience of its use in construction. The eminent scientist did his calculations, and firmly recommended that the bridge should have suspension chains for added strength. However, Fairbairn, the practical shipbuilder, was equally adamant that these would be unnecessary. He had seen the incident with the *Prince of Wales* and was converted to the cause of wrought iron.

Fairbairn experimented with different types of girder cross-sections – circular, rectangular, elliptic – to find out which was the strongest. He decided that the rectangular design was best – but this was only a starting point. At his Millwall works he built a scale model of the main tube, one-sixth of the size that would eventually be needed. This was 75 ft (23 m) between the supports, and weighed between five and six tons. Fairbairn then hung weights of increasing size from the middle until it gave way, and investigated different ways of strengthening it. The most efficient solution – that is, the one that used the least amount of iron – seemed to be to make the top plate extra thick, and to add a series of cell-like compartments to the top and bottom. So, by the application of only a ton of extra iron in the right places, the tube could bear two and a half times as much weight.

Stephenson's first plan was to have two railway tracks carried over the Menai Straits through two rectangular tubes – each of a 460-ft (140-m) span. These would be linked to each shore by stone-built viaducts similar to those used by Telford for his suspension bridge. However, as things progressed, he came up with a more revolutionary idea – to replace the stone viaducts with yet more tubes. These would be linked to the central section into one giant 1,511-ft (460-m) length.

Transporting tubes of such unprecedented size posed a problem. Eventually it was decided that the easiest solution was to make them on site. A massive timber platform was built that stretched for a thousand yards along the Caernarvon shore; altogether, there were 3¼ acres of timber staging. Iron was floated round the coast from Liverpool, and the banks of the Menai were turned into a giant blacksmith's workshop, as workmen pounded away at massive plates of iron with 40-lb (18-kg) sledgehammers. When darkness fell, the sound of the hammers and the glow of the furnaces must have seemed like Dante's Inferno.

L T C Rolt paints a vivid picture of the spectacle that would have been seen by anyone looking across the straits from Anglesey: 'Forty-eight rivet hearths pricked the length of the opposite shore line with flickering points of flame, and from them, all the night long, a golden rain of dazzling sparks shot upwards, describing graceful arcs of light to be caught and reflected in the moving seas. This dramatic effect was produced by the rivet boys who, drawing in their tongs the white-hot rivets from the hearths, hurled them 40 ft [12 m] into the air to be dextrously caught by the riveting gangs at work on the tops of the tubes.'

△ Floating a tube on the Conway Bridge. Stephenson used this as a dress rehearsal for the larger Britannia Bridge. In the background is Telford's Conway Castle Bridge.

In many ways, the story of the building of the Britannia Bridge is like a re-run of Telford's experiences a quarter of a century earlier. Stephenson's bridge was only 1 mile (1.6 km) south of Telford's suspension bridge. Like Telford, he not only lost sleep in his anxiety over the bridge in its last stages, but he obsessively rehearsed the final floating exercise. Telford had floated chains; Stephenson's rectangular tubes were infinitely heavier.

Samuel Smiles, in his account of this phase of Robert Stephenson's life, records:

> *Afterwards describing his feelings to his friend Mr [Tom] Gooch, he said: 'It was a most anxious and harassing time with me. Often at night I would lie tossing about, seeking sleep in vain. The tubes filled my head. I went to bed with them and got up with them. In the grey of the morning, when I looked across the Square [Gloucester Square, Hyde Park, where he lived] it seemed an immense distance across to the houses on the opposite side. It was nearly the same length as the span of my tubular bridge!'*

A scale model was made, with real water and real iron tubes, to practise the floating. The whole exercise was then carried out in miniature for a smaller bridge on the same railway. This was the Conway Bridge, which was completed in 1848, and had 400-ft (122-m) long tubes constructed on the same principle as the Britannia. The floating had its heart-stopping moments, but its successful completion proved the operation was possible.

The tubes for each landward end of the Britannia Bridge had been built on scaffolding in their permanent positions. The four larger tubes were built on wooden platforms at high-water mark on the Caernarvon shore. The plan was to slide pontoons under them and use the rising tide to lift them off. Once afloat, the current would grab the tubes and push them towards the stone piers. Teams of men on the shore would pull on cables tethered to the tubes to guide them to the right spot. (These cables were wound round capstans, with cable-stoppers to check the speed if necessary.) Finally they would slot each end of the tube into recesses on the piers, and wait for the tide to turn. As the water level fell, and the pontoons dropped, the tubes would come to rest on hydraulic jacks on the piers, ready to be lifted into their final positions. At least, that was the plan.

On 19 June 1849, all was ready for the floating of the first tube. The three other completed iron tubes, which were lined up along the Caernarvon shore, were converted into a grandstand for the day. Crowds waited expectantly, while Robert Stephenson conducted operations from the stand erected for the directors of the Chester and Holyhead Railway on the other side of the straits. Standing next to him were Brunel and Joseph Locke – the three giants of Victorian railway engineering all together.

The hard manual work of the floating was carried out by a team of Liverpool sailors, under the command of Captain Claxton (who had been in charge of the rescue operation on Brunel's *Great Britain*). By the early evening, the pontoons had lifted the tube on the tide, and the moment came for it to be floated off. However, part of the winding gear gave way, so the attempt had to be ignominiously abandoned. The following day was exceptionally windy; buoys were uprooted, and vital cables were whipped to and fro in a shower of spray.

When things were finally under sufficient control to consider floating, the tube took off at great speed. However, at the critical moment, there was a disaster. The end of the tube swung too far out from its allotted place at the base of the Anglesey pier; the men tried to check it, but the cable jammed on the capstan. This was then ripped from its foundations, catapulting some of the operators into the sea.

Heroically, the foreman seized the free end of the cable and pulled it away from the shore into a field, calling for help. An army of spectators grabbed hold of the cable, pitting their human muscle power against the dead weight of iron in the sea. Slowly they managed to drag the tube back until it slotted into its place; then the sailors at the other end could do their bit to complete the operation. Cannon fire and loud cheers from the watching crowd greeted this success.

This was not quite the end of the day's work, as the sailors had to get back to safety on the pontoons, battling against treacherous tides and a howling gale. However, they accomplished this safely with no casualties. That night, Stephenson slept soundly for the first time in weeks.

Happy with the outcome of the floating, Stephenson returned to London. But his experiences with the Conway Bridge had taught him to be extra cautious over the next stage. This was the lifting of the tube into place with hydraulic presses. At Conway, the second tube had been raised to within 2 ft 3 in (69 cm) of its correct height when an alarming crack appeared in part of the lifting gear. Lifting was only safely accomplished after the tube was hastily underpinned with timber packing, amid the constant danger that the whole machinery would collapse.

△ Lithograph showing the floating of the second tube on the Britannia Bridge.

Against this background – and always remembering the Dee Bridge disaster – Stephenson refused to be swept along with his workmen's enthusiasm for lifting the tube as quickly as possible. When his assistant, Edwin Clark, wrote suggesting that they could finish the task in a day or two, he insisted that they move inch by inch, building up underneath as they went along.

His caution was vindicated when, after the tube had been lifted up 24 ft (7.3 m), the cylinder of the lifting press broke. The cylinder cover and chain tackle, weighing more than fifty tons, came crashing down on to the deck of the tube, sweeping one man to his death as it fell. The iron tube fell only a few inches onto the supporting masonry, although even this was enough to cause £5,000 worth of damage. In a letter to Stephenson, Edwin Clark commented: 'Thank God you have been so obstinate. For if this accident had occurred without a bed for the tube to fall on, the whole would have been lying across the bottom of the Straits.'

On 5 March 1850, Robert Stephenson put the last rivet into the last tube and travelled by train across the bridge with a crowd of followers. The success of the Conway and Britannia bridges encouraged him to apply the same principle to several others. Two were in Egypt, and the last was the massive 6,588-ft (2,008-m) long Victoria Bridge over the River St Lawrence at Montreal.

BRITANNIA BRIDGE

ANGLESEY ENTRANCE.

The Victoria Bridge was one long tube, made up of twenty-five sections. All the ironwork was sent out to Canada from England, drilled and marked ready for assembly. However, by the time the bridge was built, it was already obsolete in technique. Travelling through its fume-filled length was a claustrophobic experience, and before half a century was out, it was rebuilt using steel lattice trusses.

In bridges, as in every other sphere of engineering, Brunel played the role of the maverick magpie. He took the best elements of other people's designs and transformed them into something that was his own. So, his development of cell-like watertight compartments for ships can be linked to Fairbairn's experiments with wrought iron. And his last bridges – at Chepstow and at Saltash – can be seen as logical extensions of Stephenson's work.

His bridge over the River Wye at Chepstow differed from the Britannia Bridge in being an airy open-sided construction, rather than a closed one. But it gained its strength from wrought-iron tubes – round, as well as rectangular. Stephenson had chosen rectangular tubes because they had

△ Lithograph by George Hawkins showing the stone lions that guard the Anglesey entrance to the Britannia Bridge.

performed the best in Fairbairn's experiments. However, this was partly because the round and oval tubes had been made too thin. Brunel carried out further experiments, and incorporated the results into his bridges.

The Chepstow Bridge had a double-line deck suspended by chains and rods from two wrought-iron tubes, carried on masonry towers. In construction and assembly, this was Brunel's equivalent of Stephenson's bridge over the River Conway – a dress rehearsal for the big set piece to come.

Just as Stephenson had invited him to be present at the floating of the Britannia Bridge tubes, so Brunel in turn wrote to his friend for moral support at Chepstow. To his chagrin, Stephenson was too tied up with parliamentary business to be present. There is a note of genuine regret in his letter of apology, written in April 1852:

▽ Stephenson's Britannia Bridge, with Telford's Menai Suspension Bridge in the background. The combination of these rail and road bridges helped to develop Holyhead as a major port for travellers to Ireland.

You have my kindest sympathy, but I think I hear you saying 'Damn the fellow's sympathy, I want none of it.'

Don't say so! I know too well what floating is not to feel the advantage of a friend alongside one in such cases, and believe me, it is a source of sincere regret that I am not with you.

At Saltash, Brunel faced a challenge very similar to Stephenson's in the Menai Straits. He had to build a bridge over the Tamar, linking the counties of Devon and Cornwall, and he had to provide sufficient headroom for ships to satisfy the Admiralty. However, unlike Stephenson, he had no handy rock in the middle to provide a stepping stone. The foundation for his central pier lay beneath 70 ft (21 m) of fast flowing water and 20 ft (6 m) of mud.

Brunel's solution was a throwback to his Thames Tunnel days. There, he had used a diving bell to examine the riverbed. Now, he proposed to custom-build his own watertight compartment in the form of a wrought-iron cylinder. The first of these was 6 ft (1.8 m) in diameter and 85 ft (26 m) in length. This was towed out and anchored in the river over the site of the proposed pier. Cables attached to barges controlled its descent, and water was pumped out so it could act like a coffer dam on the riverbed. (A coffer dam, more usually created out of piles with clay packed between them, is a watertight enclosure that creates a dry space for site investigations, or for building foundations.)

William Glennie, who had been the Resident Engineer on the Box Tunnel, was in charge of operations. By 25 May 1848 he reported to Brunel that he had been able to stand in the mud at the bottom of the Tamar and give it a close examination:

It is most favourable for our operations being very adhesive and stiff and perfectly watertight. We had to dig it with a spade. It contains a small portion of sand and broken shells but no gravel as far as we have examined it.

He also noted that he had smoked a cigar at the bottom of the river, a theatrical gesture Brunel would no doubt have appreciated.

Investigations using the cylinder continued throughout the autumn and into the cold of winter. Altogether, the apparatus was moved along the riverbed thirty-five times, with a detailed examination being made of each new section. This meant that a model could be made of the riverbed, to determine the best place to sink the pier. Success at keeping the cylinder watertight also meant that Brunel now had the confidence he needed for the next stage of his great plan.

Progress was hampered by difficulty in raising money for the project. Railway Mania had subsided; people no longer saw investment in iron roads as an easy way of making a fortune. For three years, the only action was on

△ Brunel's bridge over the River Wye at Chepstow. This was built for double track working, unlike the Saltash bridge, where Brunel had to compromise to cut costs.

paper as Brunel tried to work out how to cut costs. In 1852, he wrote to the Board of Trade with his solution:

> *This bridge had been always assumed to be constructed for a double line of railway as well as the rest of the line. In constructing the whole of the line at present with a single line of rails, except at certain places, the prospect of doubling it hereafter is not wholly abandoned, but with respect to the bridge it is otherwise.*
>
> *It is now universally admitted that when a sufficient object is to be attained, arrangements may easily be made by which a short piece of single line can be worked without any appreciable inconvenience... This will make a reduction of at least £100,000.*

The bridge was therefore constructed for single-line working, and so it has remained until the present day. Compromise was not Brunel's middle name as a rule, and the fact that he was prepared to do this suggests he realized that it was half a bridge or nothing. Either that or he was beginning to be aware that his time was short. For although he was only in his late forties, he had lived his life at double speed, and his health was starting to give out.

By February 1853, he was able to report that the contract for the building of the bridge had been let to the same Blackwall shipbuilding firm that had been responsible for the construction of the Britannia Bridge. They moved in to a site on the Devon shores of the Tamar, converting a quiet backwater known as Saltash Passage into a massive ironworks. On 2 September, the *West Briton and Cornwall Advertiser* reported:

> *Extensive workshops and smithies have been erected. Steam machinery of every description for planing, rolling into shape, cutting, drilling, and punching the masses of iron to form parts of the bridge, is in full operation. The smiths' shop contains eight forges, worked upon the principle of exhaustion, of what is commonly called fan bellows, driven by steam.*

△ The Great Cylinder being floated out. Divers became sick when working in this contraption.

Brunel's masterstroke for the construction of the central pier was the Great Cylinder. This had a shell made from wrought-iron boilerplates, approximately 37 ft (11 m) in diameter and 90 ft (27.5 m) in overall length. One end was shaped to fit the slope of the riverbed, with one side being 6 ft (1.8 m) longer than the other. Including all its internal fittings, it weighed some three hundred tons.

The cylinder had been designed as a tube within a tube. The inner tube was a diving bell within which the masons worked. In the original scheme of things, this was to be kept clear of water by means of pumps and ventilated by a pipe that extended from the top of the dome of the bell to the full height of the outer cylinder. Only the space between the mason's working chamber and the outside of the cylinder was to be sealed and pressurized. The workmen who cleared the mud in this outer area had to reach it through air-lock doors.

By February 1855, the cylinder had been floated out and lowered into place. However, dense beds of oyster shells, which had to be excavated, caused some delay. This was compounded by the hardness of the greenstone trap rock, and the discovery of an unexpected fissure, which forced water like a jet-propelled fountain into the masons' inner compartment.

Brunel's hopes of being able to avoid the cumbersome business of passing men and materials through air-lock doors into the inner chamber were dashed. This now had to be pressurized in the same way as the outer

chamber before work could continue. It also had to be strengthened, to prevent the extra air pressure from bursting its sides, and loaded up with 750 tons of ballast to prevent it from popping up to the surface like a cork. Amazingly, all this was done while the cylinder was still in the middle of the river.

Being locked into the underwater gloom, working in an eerie silence broken only by the sound of their tools, must have been hard. The men's initial task was the back-breaking one of hacking away at rock and clearing mud to gain a purchase for the masonry of the pier. Levelling the rock in some places meant going as far down as 6 ft (1.8 m) under the riverbed, every inch being painstakingly nibbled away with hand chisels. It would have been damp and cold, with the only light provided by miner's tallow candles. And, as if this was not enough, there was a problem called decompression sickness that divers the world over now know as 'the bends'.

Very little was known at this time about decompression sickness, a condition that develops after working at depth under water. When men who had been toiling on the riverbed suddenly returned to the normal pressure at the surface, dissolved gases in their blood started to form bubbles,

which often caused blockages in their blood vessels. This was a painful and potentially fatal condition. It was only when the men became sick after working seven-hour shifts underwater that anyone realized there was a problem. Shortening the shifts seemed to help, but there is no record of any other precautions being taken.

The cylinder had been designed so that its lower part could be filled with masonry and left in place underwater. The upper part was constructed in two pieces of larger diameter, bolted together, so it could be removed and floated back to shore. By the end of November 1856, the massive granite centre pier had been completed and the Great Cylinder returned to shore. However, success, had come at the cost of bankruptcy for the principal contractor, Mare and Co. Sadly this was by no means an isolated example of an experienced contractor falling victim to the unexpected cost of Brunel's ambitious plans.

Meanwhile the main company took over the rest of the construction work. The two spans of the bridge were designed in the form of a bowstring suspension bridge. Each span had a wrought-iron tubular arch, or bow, with sets of suspension chains hanging down. A massive plate girder for the rails was slung below the tube from eleven pairs of supports, which were connected to the chains. Conventional wisdom of the time had it that suspension bridges were unsuitable for railways, so Brunel provided a complicated network of extra stays and supports.

The two main tubes were oval in section. This meant that their breadth could be the same as that of the roadway below, so the suspended chains could hang vertically (unlike the Chepstow Bridge, where they hung at an angle). The chains that were used had, in fact, originally been intended for the Clifton Suspension Bridge, a sad reminder of another equally ambitious project that had been scuppered by lack of funds.

Floating the tubes was no small undertaking. Each one weighed over a thousand tons, and even though Brunel had the experience of Stephenson's Britannia Bridge to help him, he was not going to risk failure. Everything was meticulously planned in his notebooks, including the flag signals to be used:

> *Signals by flag:*
> *Heave in – red.*
> *Hold on – white.*
> *Pay out – blue.*
> *Waved gently means gently.*
> *Waved violently means quickly.*

△ Royal Albert Bridge at Saltash. The Cornwall span in position; the Devon span still under construction. Spring 1858.

The floating of the first span, on 1 September 1857, was conducted with the precision of a military operation, albeit a somewhat theatrical one. Brunel had a platform specially built so he could co-ordinate the movements of some five hundred men on shore and afloat. He insisted on silence until the ends of the tube were securely in place. Only then did the cheering break out from the assembled crowd, as the band of the Royal Marines struck up with 'See the Conquering Hero Comes'.

Isambard 'Barnum' Brunel had done it again. However, what neither he nor the crowd knew was that, when the bridge was finally completed, he would not be in a position to conduct operations from anything other than an invalid's couch.

THE GREAT EASTERN

The Great Exhibition of 1851 firmly enshrined Brunel's and Stephenson's places in the pantheon of Victorian engineering. Not only did they both help to organize the whole event, but their projects filled the aisles of the giant greenhouse built to house it.

Joseph Paxton's iron-and-glass exhibition building, which became known as the 'Crystal Palace', was the perfect backdrop for the grandest showpieces of the age. The sheer scale of it, with 293,655 panes of glass and 24 miles (38 km) of guttering, was just the kind of megalithic scheme that could have come from Brunel's brain. (In fact, as we have seen, he had originally produced his own design for the building, which had little to recommend it but its size.)

Brunel walked fourth from the front in the grand opening procession led by Queen Victoria and Prince Albert on May Day. Some twenty-five thousand people visited on the first day. By the time the exhibition closed in October, over six million had paid their one shilling (5p) to wander up and down the aisles in awe. Many had travelled on special excursion trains laid on by the big railway companies; when they got there, they found that railways loomed large among the exhibits, too.

Heavy machinery of all kinds – especially working machinery – was a big attraction. One section, labelled 'Machines for Direct Use', was dominated by a Great Western Railway broad-gauge steam engine, the 31-ton *Lord of the Isles*. Brunel's Chepstow Bridge made an appearance in the form of an uncharacteristically small model. In the main avenue was a larger model of Stephenson's Britannia Bridge; elsewhere the contractors exhibited the hydraulic press and crane that had been used to build it.

One of the wackier exhibits was an eighty-blade sportsman's knife, manufactured by Joseph Rodgers and Sons of Sheffield. This device, like a massively elaborate but totally impractical Swiss Army knife, featured a gold-inlaid handle with representations of the icons of the time. These included the Crystal Palace, Windsor Castle – and the Britannia Bridge.

By this stage of his career, Stephenson was content simply to continue to do what he knew he could do well. For Brunel, the very fact that he could do something was an incentive never to repeat it. His restless ambition was always looking for something bigger and better – and, around the time of the Great Exhibition, his mind turned once again to paddle wheels and screw propellers.

It might be thought that Brunel would have had enough of ships, with all the difficulties he had encountered with the *Great Western* and the *Great Britain*. However, in 1851, the Australian Mail Company asked his advice on the best size and class of ship to operate a mail service from Britain to Australia. This was a challenge bigger than any Brunel had yet attempted, and he was hooked.

'Brunel comes back into the ship business because a new opportunity opens up,' explains Andrew Lambert. 'He doesn't need to go back onto the North Atlantic – he's done that, he's proved his point. Nobody does the North Atlantic better than Brunel. The new route is to Australia. Australia has discovered gold; there's a new market, a new opportunity, and this isn't 3,000 miles [4,800 km], this is 12,000 miles [19,300 km] – this is vast. But the demand is there, the interest is there, and with demand and interest you can raise the capital to get another project going.'

Gold-diggers and farmers alike were attracted to Australia. The Great Exhibition had shown the extent of Australia's agricultural riches, and many who struggled to make a living on the land in the 'hungry forties' were keen to start a new life elsewhere. Brunel's *Great Britain* had, in fact, been put on the Australia run after it had been repaired. However, despite its massive size, it was still not big enough to take its own coal around the world. This had to be transported in bunker ships, so it could refuel in South Africa.

△ Previous page: Robert Stephenson in the 1850s. Oil portrait by John Lucas.

▽ Joseph Paxton's Crystal Palace was tall enough to be built around trees already standing in Hyde Park. Exhibits of machinery in motion were popular, with attendants and printed cards to explain how things worked.

Brunel teamed up with a London shipbuilder, John Scott Russell, on a couple of iron steamships for the Australian run, the *Adelaide* and the *Victoria*. The design for these ships, which were constructed at Russell's Millwall shipyard, was Russell's responsibility rather than Brunel's. They had their limitations; they still had to stop at the Cape to take on coal. However, out of them came the project known in the early days as the 'Great Ship', or the 'Leviathan', and eventually as the *Great Eastern*.

According to Andrew Lambert, the scheme was built around a faulty premise: that there was no coal in Australia. In fact, Australia did have coal – appropriately enough, at a place called Newcastle. However, the *Great*

△ The paddle engine room in the *Great Eastern.*

Western had developed steaming problems caused by poor-quality coal supplied locally in New York, and Brunel would not have wanted to risk a repeat of this. Shipping British coal separately brought with it the possibility of delay to the operating schedules, so the idea of creating a massive floating coal bunker that could also carry passengers had its attractions.

Leaving aside the economics of the project, there is the sense that Brunel simply seized this chance to build the biggest ship ever. The *Great Eastern* was to be over three times the length of the *Great Britain*, and capable of carrying 4,000 passengers. Only the *Lusitania*, launched forty-eight years later, was to exceed it in tonnage.

'It's far heavier engineering than Brunel used in the *Great Britain*,' explains Andrew Lambert. 'He's gone from the elegant lightness of the *Great Britain* to a kind of mid-Victorian monumentality. The *Great Eastern* is held together with rivets bigger than any other ship; its frames and beams and members are more colossal than anything else. It's a project for an older man, a man who is thinking not of risk and speed, but more of comfort and dependability. This ship will get through. It's interesting that, fifty years later, she bankrupted the first three companies that tried to break her up. She was built so well that she almost couldn't be taken to pieces – it's a shame in many ways that she was.'

The *Great Eastern* was the prototype for all modern ocean liners. She had a double-bottomed hull, with watertight compartments, so a puncture in the skin of the ship would not sink it. The original plan had been for an inner hull of iron, with an outer skin of copper-sheathed wood, but she was eventually built with both skins being made of iron. Her top deck was made up of many hundreds of cells, to strengthen it, just like Stephenson's Britannia Bridge. All this made her virtually unsinkable.

She was powered both by paddle wheels and by a screw propeller. It has been suggested that what looks like engineering overkill was simply

▽ Photographs of human figures standing by various parts of the *Great Eastern* were a popular way of showing its scale. This picture shows the ship's stern, with propeller.

due to the difficulties of making a large enough engine, propeller and propeller shaft. However, Brunel was not usually intimidated by size; there were other advantages to the dual system.

'Remember, this is a monstrous ship,' says Andrew Lambert. 'The existing harbours and docks are not designed to deal with ships of this size, and it's going to be very difficult to manoeuvre. So the paddle wheels, which can run in opposite directions, allow the ship to manoeuvre very well and actually turn out to be very useful when she goes into another existence after Brunel's death.'

Innovative in its size and its use of watertight compartments, the *Great Eastern* was also innovative in its style. Its stark, functional lines made it seem much more modern in design than the *Great Britain*, which featured elegant flourishes and stern galleries like an old-fashioned man of war. As Andrew Lambert puts it:

'The *Great Eastern* has nothing. It rises straight up out of the sea and that's it – not even a decorative flourish at the end. This is the aesthetic of the age. It's not harking back, it's looking forward – it's an industrial monument. The *Great Britain* pretended to be a wooden sailing ship; the *Great Eastern* pretended to be a railway station.'

Brunel was responsible for the design of the ship and for detailing the specifications. The contract for constructing the hull was given to John Scott Russell, who had tendered a bid so low that it should have alerted Brunel to the possibility of troubles to come. However, as always, Isambard had numerous projects running in parallel.

At the time the contract was signed, in December 1853, Brunel was engineer to seven railways, including the Great Western. The finishing touches were being put to Paddington Station, and the Royal Albert Bridge at Saltash was still under construction. He was also preparing to take over the design of two water towers to service the network of fountains at the re-erected Crystal Palace in Sydenham. Such intensive activity was nothing new; neither was his insistence on maintaining an iron grip on the smallest detail. However, the deterioration in his health brought an increasing note of tetchiness in his dealings with others.

When the directors of the Eastern Steam Navigation Company – understandably anxious about his workload – asked him to nominate a Resident Engineer, his reply was breathless with fury. He wrote:

I cannot act under any supervision, or form part of any system which recognises any other adviser than myself, or from any other source of

△ John Scott Russell, who became involved in a power struggle with Brunel. Portrait by Henry Wyndham Phillips.

information than mine, on any question connected with the construction or mode of carrying out practically this great project on which I have staked my character; nor could I continue to act if it could be assumed for a moment that the work required to be looked after by a Director, or anybody but myself or those employed directly by me and for me personally for that purpose. If any doubt ever arises on these points I must cease to be responsible and cease to act.

The question of control surfaced with a vengeance in his relationship with John Scott Russell. Two years younger than Brunel, Scott Russell was a student of Edinburgh and Glasgow Universities, and had been a mathematical prodigy. He designed his ship hulls according to the wave-line theory – a scientific principle that he himself had developed. The idea was to use mathematical calculations to shape the body of the ship so it would glide through the water in harmony with the waves.

He and Brunel had become close when they both worked on the Great Exhibition, but their friendship was not to survive their collaboration on the *Great Eastern*. Brunel considered himself both the originator of the project and the designer of the ship, and took grave offence when he felt he was not being given due credit. In November 1854, the *Observer* published an article entitled 'Iron Steam Ships – the Leviathan'. His name was mentioned only once: 'Mr Brunel, the Engineer of the Eastern Steam Navigation Company, approved of the project, and Mr Scott Russell undertook to carry out the design.' The article was unsigned, but it read as though it had been prompted by an insider. This impression was reinforced by the fact that John Yates, the company secretary, had been distributing copies in apparent approval of the contents.

Brunel immediately jumped to the conclusion that it was part of a campaign by either Yates or Scott Russell or both to relegate him to the background. He wrote in high dudgeon to Yates:

I cannot allow it to be stated, apparently on authority, while I have the whole heavy responsibility of its success resting on my shoulders, that I

*am a mere passive approver of the project of another, which in fact
originated solely with me, and has been worked out by me at great cost of
labour and thought devoted to it now for not less than three years.*

The ambition of Brunel's plans for the ship can be seen in his notebooks. In
October 1855, under the heading 'Memoranda for my Own Guidance', he
listed a range of considerations, numbered one to twelve. These started
with the provision of a large icehouse, and a railway on deck to carry dinner
from the kitchen to each saloon. Number eleven was the idea that the
chimneys should be oval – and number twelve, a system of emergency
steering by semaphore signals, with a loud bell to call attention.

Navigational details were worked out in a long and complicated
correspondence with Professor Airy, the Astronomer Royal, and Professor
Piazzi Smyth, the Astronomer Royal for Scotland. Brunel's idea was that,
as the size of the ship would make it exceptionally stable, a gyroscope
could be used as a stand for astronomical instruments. He also intended that

△ *Building the steamship Great Eastern, by William Parrot.*

a constant stream of water would be pumped up through the observers' cabin, and its temperature measured to indicate the presence of icebergs. (The normal way was for the occasional bucketful of water to be hauled up by the watch officer on deck.)

Some of his design ideas never made it beyond the drawing board, as the project ran into money problems at an early stage. The original plan had been to build the ship in a dry dock. However, it soon became obvious that it would be prohibitively expensive to construct a dock big enough. Accordingly, the work was carried out at two shipyards in Millwall: one belonging to Scott Russell, the other an empty one next door to it, specially leased for the purpose. Two 600-ton cargoes of timber were used to create a cradle of oak piles for the ship, which was to be launched sideways into the Thames. The piles were then strengthened with concrete, and iron rails were laid on the launching path.

The sideways launch had been decided upon because of the ship's great size, and the angle of the slipway. If she had been built for a conventional lengthways launch, the highest part of the hull would have been 100 ft (30 m) off the ground, posing immense problems for the workmen. The launch would also have put a severe strain on the structure.

From the outset, it was clear that the mechanics of manoeuvring this 18,915-ton whale of a ship into the river were going to be problematic. As the giant structure took shape on the bank, it became the subject of much public speculation, and Brunel was certainly not short of advice. A Mr G W Bull from Buffalo wrote to say that hundreds of boats had been launched sideways into Lake Erie without mishap. However, these had been free launches, and what Brunel proposed was an infinitely more expensive controlled launch, with the ship's progress being checked by chains. Russell thought this unnecessary, and this was just one of many disagreements between the two men.

Building the ship was a labour-intensive business, with every one of the millions of rivets being closed by hand. Each iron plate was lifted from the ground with fairly primitive lifting tackle; Russell had wanted to build a crane, at a cost of almost £3,000 but Brunel balked at this figure. Even so, Russell had to invest in new plate-bending and punching machines, and to sink coffer dams into his foundry floor to enable the 40-ft (12-m) high paddle engine cylinders to be cast. The crankshaft was contracted out, after some difficulty, to a forge in Glasgow. The shaft for the *Great Britain* – which had precipitated the invention of James Nasmyth's steam hammer – had been daunting enough. But, as L T C Rolt puts it, that was like a bent pin to a crowbar compared to this.

Predictably, Russell soon ran into cash-flow problems. His bankers refused to grant him further credit, and Brunel had to authorize extra payments. However, he became impatient at the slow rate of progress, and what he saw as lack of co-operation. The tone of their correspondence became increasingly acrimonious, Scott Russell's suave politeness only serving to incense Brunel further. In October 1855, furious at the vagueness of an answer Russell had sent about the projected weight of the ship, Brunel wrote:

> *How the devil can you say you satisfied yourself of the weight of the ship when the figures your Clerk gave you are 1,000 tons less than I make it or than you made it a few months ago? For __shame__, if you are satisfied. I am sorry to give you trouble but I think you will thank me for it. I wish you __were__ my obedient servant, I should begin by a little flogging.*

By the New Year, Brunel was also becoming suspicious about the apparent disappearance of large quantities of iron from the shipyard. This letter had a clear accusation of dishonesty concealed beneath a veneer of politeness:

> *My Dear Russell,*
> *It is impossible that I can feel otherwise than greatly alarmed at the appearance of the state of things as regards 'stock'. There are 2,400 tons to be accounted for. If my fears should prove too well founded let me entreat you as a sincere friend to meet the thing openly and to trace up the explanation and to give it plainly.*

With friends like this, who needs enemies? It is hard at this distance to work out exactly who was in the right in the dispute between the two men. L T C Rolt, in his biography of Brunel, makes detailed allegations of financial mismanagement, and also implies that a fire in Russell's shipyard was started deliberately. Adrian Vaughan, on the other hand, says that Russell's personal honesty and engineering reputation has been 'shamefully attacked' by Rolt. The truth probably lies somewhere in between – but it seems likely that the root of the rift between the two men was jealousy. As Andrew Lambert puts it:

'Scott Russell puts himself in harm's way because he's not accounting properly for what he's doing, he's not overseeing materials, he's not managing the men, the money or the materials, the ship is in trouble and the funds to build her are running out. Essentially, he's got his hand in the till and Brunel really, I think, knows this and he doesn't know how to deal with it. He's desperate to finish the job, get it out of Scott Russell's hands and get it away. This just doesn't happen – over and over again, delays occur.

'Scott Russell then does the unpardonable and claims more and more that the credit for the ship is of course his, and that he designed the *Great Eastern*. Aside from a few flow lines on the hull it's very difficult to see what Scott Russell could have contributed. This infuriates Brunel I think more even than having his hand in the till. This man's trying to steal his glory; the money is a trifle – the glory is the important thing. So they fall out monumentally, but Scott Russell is a skilled survivor and operator and he carries on. He managed to keep himself involved with the *Great Eastern* until after Brunel's death.

'The core of the relationship between Brunel and Scott Russell is the one we find between all prima donnas. Both these men want to be the star of this project; they want their name in lights above the name of the ship. The analogy would be with a modern film: who gets his name above the credits and above the title, who gets their name underneath. It's Isambard Kingdom Brunel's *Great Eastern*, also starring John Scott Russell. John Scott Russell doesn't want that. He wants either joint top billing or sole top billing, with Brunel relegated to creative consultant, art director – something along those lines. That's what they're fighting about.

'So when Scott Russell says, "It's my ship", he's not claiming just a credit for the design, he's claiming the kudos, the celebrity, the notoriety of being the man behind the great ship. Because it is *the* great ship. Nobody bothers to call it by its name – it doesn't need a name, it's the obvious. They fight over that and that ultimately is what tears their relationship apart and destroys Brunel. Scott Russell is quite happy working in this world. He's done this before – he's prepared to fight nasty for these things. Brunel I don't think is well enough, or prepared to fight in this way, and I think that's one of the things that saps his strength.'

Because the *Great Eastern* loomed so large in the closing stages of Brunel's career, it is easy to forget that – as usual – he had many other concerns. War with Russia, which had begun in March 1854, provided a distraction for many in the scientific and engineering establishments. Brunel's contribution was to design a floating gun carriage, which was never built, and a prefabricated hospital, which was.

The newspaper reports of W H Russell in *The Times*, and the efforts of Florence Nightingale, had done much to alert the British public to the appalling conditions facing soldiers in the Crimean War. Out of a total of 56,000 troops, 34,000 died between September 1854 and January 1855. Hospitals were hopelessly overcrowded and understaffed; more died from epidemics arising from the insanitary conditions than on the battlefield.

Brunel's brother-in-law, Benjamin Hawes, was Permanent Under Secretary at the War Office. In February 1855, after the allied armies had suffered their first disastrous winter in the Crimea, Hawes wrote to Brunel asking if he would be prepared to try his hand at hospital design. The idea would be to come up with a prefabricated building that could be shipped out from England, and assembled wherever it was needed. Brunel replied on the same day that he received the letter, saying that his 'time and best exertions would be, without any limitations, entirely at the service of Government'.

Only six days later, he was writing to Hawes with a request for contoured sketch maps of suggested sites. He asked around for advice, and even had an experimental ward erected at Paddington so people could comment. Unusually for him, Brunel also had a paper printed setting out his ideas. According to his son, this was written 'to satisfy the curiosity of his friends'. However, it could also have been excitement at exploring new territory once more, mixed with a growing need to explain himself for posterity:

> *March 1855.*
>
> *The conditions that it was considered necessary to lay down in designing these buildings were —*
>
> *First. That they should be capable of adapting themselves to any plot of ground that might be selected, whatever its form, level or inclination, within reasonable limits.*
>
> *Secondly. That each set of buildings should be capable of being easily extended from one holding 500 patients to one for 1,000 or 1,500, or whatever might be the limit which sanitary or other conditions might prescribe.*
>
> *Thirdly. That when erected they might be sure to contain every comfort which it would be possible under the circumstances to afford. And —*
>
> *Fourthly. That they should be very portable, and of the cheapest construction.*

The idea of prefabricated buildings was not new, but the flexibility built into Brunel's design was. Essentially, he provided a modular system

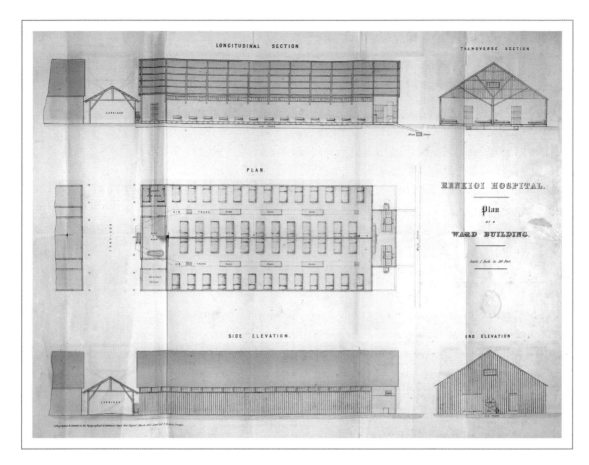

LONGITUDINAL SECTION.

TRANSVERSE SECTION

PLAN.

RENKIOI HOSPITAL.

Plan
of a
WARD BUILDING.

SIDE ELEVATION.

END ELEVATION

△ Plan of ward building for Renkioi Hospital. Each building contained two wards, designed to take 26 beds each, and was provided with ventilation. The outside walls were whitewashed to help deflect the glare of the sun.

(although he would not have used that expression). Huts made of wood and tin, each of the same size and shape, could be arranged as necessary, connected by an indefinite length of open corridor, which was boarded up in winter. An iron kitchen and washhouse would be sited a little away from the other buildings, because of the fire risk. Every detail of heating, drainage, ventilation and sanitation was thought through, along with the practicalities of transporting this portable hospital to its site. Everything had to be arranged in packages light enough to be carried by two men, and every contingency planned for:

> *For the transport of the materials to the spot selected, two sailing-vessels and three steamboats, capable of carrying one hospital for 1,000 men, which is the first about to be sent out, have been secured. In each vessel is set a certain number of complete buildings, with every detail, including their proportion of water-pipes and drains, closets, lavatories, baths, &c., and a small amount of surplus material and tools; and in each of two separate*

*vessels are sent a set of pumps and mains, and a kitchen and washhouse. So
that by no accident, mistake or confusion, short of the loss of several of the
ships, can there fail to be a certain amount of hospital accommodation,
provided with every comfort and essential.*

From the outset, Brunel was well aware of the need to move swiftly. Within
days of starting work on the project, he had placed an initial contract for
the supply of buildings. Such hastiness was hardly regulation procedure, but
as he explained to the War Contracts Department: 'Such buildings if wanted
at all, must be wanted before they can possibly arrive.' The first shipment
of materials and stores arrived in the Crimea on 7 May – less than three
months after Hawes's first letter to Brunel.

Reluctantly accepting that he could not supervise its erection himself, Brunel
chose an assistant engineer, John Brunton. John was the son of William
Brunton, who had once been a rival for the post of GWR Engineer, and was
clearly a man Brunel felt he could trust. John Brunton went out to the Crimea
to choose a suitable site, in collaboration with Dr Edmund Parkes, a civilian
doctor who had been appointed Medical Superintendent of the new hospital.

Initially, Scutari – where there was already an army hospital, situated in
a converted Turkish barracks – was considered. However, there proved to
be no suitable ground at this location. It was eventually decided that Renkioi,
a malaria-free village on a hilltop, would be healthier, albeit further from
the action. As Brunton and Parkes settled in, Brunel sent out lists of the
contents of each shipment, together with detailed instructions. These covered
everything, right down to the lavatory paper. As he wrote to Dr Parkes:

*You will be amazed to find also certain boxes of paper for the water closets – I
find that at a cost of a few shillings per day an ample supply could be
furnished and the mechanical success of the WCs will be much influenced by
this. I hope you will succeed in getting it used and not abused. In order to
assist in this important object I send out some printed notices or handbills to
be stuck up, if you see no objection, in the closet room opposite each closet
exhorting the men to use the apparatus properly and telling them how to do so.
If you do not approve of such appeals the paper can be used for other purposes
and perhaps impart some information in its exit from this upper world.*

Building took longer than expected, owing to the difficulty of finding local
labour. In the end, the entire hospital was built by the eighteen men who had

been shipped out from England. By 12 July, the hospital was ready to receive its first patients; however, these did not arrive until October. From then until February 1856, 1,331 wounded soldiers were admitted, of whom fifty died. At Scutari, the peak mortality rate had been over 42 per cent, so this figure was a considerable vindication of the healthier atmosphere at Renkioi. It should be remembered, though, that the hospital was never used to its full capacity, and those casualties it took in tended to be less acutely ill than those at Scutari.

Plans were in hand to build a railway line underneath the covered way, to deposit patients at the doors of their wards, when peace broke out. By the middle of July 1856, the hospital had been cleared of staff and patients, and by September the buildings had all been sold. Some buildings were used to house homeless victims of a fire in Salonica. The treasured water closets were sold on to a military hospital in Southampton.

Although Renkioi Hospital was in use for a relatively short time, the speed and efficiency of its construction proved what could be done simply by working things out from first principles. Brunel had never designed hospitals before, neither was he an expert on public health. However, he knew where to go for advice – and, most importantly, he was not hamstrung by what other people had done before him.

Closer to home, his other great project was that essential backdrop for the wealthy gentleman of fashion – a country estate. Brunel's home in London – where he also had his office – was a showpiece of heavy Victorian taste. He had bought 18 Duke Street in 1836, and extended into the house next door twelve years later. An inventory of the household, taken at a time when financial disaster threatened over the *Great Eastern*, suggests some elaborate conspicuous consumption, oddly at variance with the relative simplicity of Brunel's usual taste in design. Perhaps the chandeliers, candelabra, Dresden china, crimson silk curtains and richly carved furniture (which alone was valued at £4,607) were more to Mary's taste than his. However, the dining room paintings featuring Shakespearean scenes were definitely Brunel's idea. They were specially commissioned from eleven of the most famous artists of the day.

In the summer of 1847, while he was working on the South Devon Railway, Brunel had rented a furnished house in Torquay. This gave him the idea of building a country retreat in the area, and he bought some land at Watcombe, 3 miles (5 km) north of the coast – a spot where he had frequently stopped to admire the view. Every summer afterwards, when his work would allow him, he visited the area with his family. His notebooks and letters are full of sketches and ideas for his mansion and associated landscape gardens. In the event, only

△ Architect's drawing for Watcombe Park. A country mansion was the ultimate ambition for the prosperous Victorian gentleman, but Brunel's ambitious Italianate villa was never built, although the gardens were laid out.

the gardens were laid out; the money was never found to build the house, and Brunel's health gave out before he could realize this part of his ambition.

Robert Stephenson's plans for retirement involved running away to sea, rather than building houses. As he wrote to a friend, 'Ships have no knockers, happily'. His prominence as an engineer, together with his position as an MP, put him constantly at the mercy of people wanting him to promote their schemes. His biographer, J C Jeaffreson, described a typical scene:

When casual indisposition kept the engineer from Great George Street [where he had his office], *and confined him to his house, swarms of talkative, and for the most part profitless clients intruded on the privacy of the man whose too pliant temper laid him open to their annoyance. An intimate friend and colleague of Robert Stephenson says that, calling in at Gloucester Square to consult the master of the house on urgent business, he found every reception-room occupied by a crowd of persons. Being much engaged, and wishing to employ his time with correspondence till he could have an interview with Robert Stephenson, the narrator asked the servant to show him into a room where he could be by himself, and write his letters in quiet. 'If you want that, sir,' the man answered, 'you must go upstairs into one of the bedrooms, for every sitting-room is occupied with gentlemen who insist on seeing Mr Stephenson,*

although they know he is unwell.' And the caller, acting on the advice, went
upstairs and sat in a bedroom, till he could be admitted into the library.

Stephenson's main relaxation was riding – either in the London parks, or in what was then the countryside of Hampstead. However, in private life, he was a lonely widower. He had never recovered from the early death of his wife, Fanny, and had no heart for investing in grand country houses. What spare money he had went on building and furnishing his yacht, *Titania*. In fact, he had had two yachts of this name, the second being built when the first one was destroyed by fire; both were designed and constructed by John Scott Russell.

The *Titania* had a 16-ft-by-15-ft (4.9-m-by-4.6-m) saloon, large sleeping cabins, and every possible shore-side comfort. There was a crew of sixteen, a cook, and a well-stocked cellar. Passengers selected books from the ample library before turning into their bunks at night, and every Sunday at sea, Stephenson read the Church of England form of service.

Unlike Brunel, Stephenson never had any ambition to design ships, however much he enjoyed sailing in them. He was happy to leave the details to the experts. This, as has been amply demonstrated, was never Brunel's way.

Progress on Brunel's giant *Great Eastern* had been erratic, coming to a complete standstill for three months in 1856, when Scott Russell faced near bankruptcy. It was only after weeks of argument with Russell's creditors that the Great Eastern Steamship Navigation Company was allowed to move in and continue work. Even then, it was under constant pressure from bankers to finish speedily.

Meanwhile, the company tried to get some of its money back by selling tickets to sightseers. Brunel should not have been surprised at the level of public interest in this gargantuan project, but as the launch drew near, it was the last thing he wanted. In a letter to the company directors in October 1857, he attempted to make the whole process sound as risk-free and uninteresting as possible. He wrote:

The ship will not be 'launched', in the ordinary sense of the term, but merely
lowered or drawn down to low-water mark, to be thence floated off by a slow
and laborious operation, requiring two and possibly three tides, and very
probably effected partly in the night, and at no one time offering any
particularly interesting spectacle, or even the excitement of risk; as I am
happy to feel that, even assuming accidents to occur or miscalculations to
have been made, rendering the operation unsuccessful – the ship may stop

halfway or not move at all, more power or other remedies may have to be
applied – but no injury to the ship can result from any failure in the course
of proceeding in this mode of launching.

This statement, like a textbook exercise in How Not To Write a Press Release, was published in all the newspapers. However, neither its rambling sentences, nor its low-key language had any effect whatsoever in downplaying the significance of the coming launch. To make things worse, Russell's mortgagees moved in to take possession of the shipyard. This forced Brunel into launching the ship before he had had time to test the tackle properly.

Early on a raw November morning, workmen moved in to rub the launching rails with a mixture of oil and black lead. All the shores and props that had been supporting the weight of the ship had been removed, and she now rested entirely on her cradles. Brunel was so determined to have a quiet launch, he had not even invited Robert Stephenson. His idea was for the whole operation to be performed in complete silence. However, his plans were scuppered by the company directors. Without telling him, they had sold over three thousand tickets giving admission to the yard. Isambard Brunel, in his biography of his father, recorded:

> *A few days before, Mr Brunel had suggested that four policemen should be*
> *obtained, thinking that all they would have to do would be to contend with*
> *trespassers. The police force actually present were ignorant of the portions of*
> *the yard to be kept clear, and Mr Brunel had himself to go and assist in*
> *ordering visitors away from the neighbourhood of the path prepared for the*
> *tackle of the stern hauling gear. The crowd soon became so great that it was*
> *almost impossible for the men in charge of the hauling-engine at the stern to*
> *see the signals given from the middle of the yard, or for those in the middle of*
> *the yard to see what was happening at the stern.*

The same signalling arrangements were used as for the floating of the first truss of the bridge at Saltash two months before. However, there the similarities ended. Saltash had been a controlled, disciplined affair; this was a fiasco.

The ship's cradles were attached to heavy chains, which were wound around giant checking drums. When the order was given, men turned the gearing handles attached to the drums to pay out the chain. Shortly afterwards, winches mounted on moored barges and on shore were started up. As little seemed to be happening, apart from a lot of thundering and

△ Watching the failed launch of the *Great Eastern*, November 1857. Brunel is second from the right.

clanking, Brunel signalled for the hydraulic presses to be brought into play. The ship lurched forward with an almighty jerk, taking up the slack of the chains, and leaving the handles on the checking drums spinning wildly. An Irish labourer named Donovan, who was unfortunately in the way, had his legs smashed to pieces by a flailing handle. He died shortly afterwards, and four other men were also injured.

By the time the ship was brought under control, using the brakes on the checking drums, her bows had moved 3 ft (90 cm), and her stern just over 4 ft (122 cm). After a pause to regroup, another attempt was made to restart the launch. This time, Brunel decided not to use the barges moored in the river. At the first launch attempt, one barge man had jumped ship, pushing off in a small boat and leaving the others to their fate. Not surprisingly, he had been less than enthralled at the prospect of being mown down by an iron hulk weighing almost twenty thousand tons.

The afternoon's launch attempt went little better than the morning's, when some of the machinery broke. Brunel's investigations were hampered by the crowds in the yard, so there was little to be done but wait until the following day. *The Times* reported sarcastically: 'We seem to have been a little unfortunate in our grandiose schemes of late.' As Andrew Lambert explains:

△ The *Great Eastern*, with one of its giant checking drums. This photograph, like the one on the previous page, is one of a series taken by Robert Howlett.

'They're not paying to see the ship launched, they're paying to see a disaster, and the disaster they get is not the one they're expecting. It doesn't career down the riverbank, it doesn't rush into the river and drown people and knock people off the staging – it doesn't do anything. So they're very displeased – they've paid good money and they've seen nothing. It's a disaster, but not the kind of disaster they were looking for.

'Then, as much as today, we have this love-hate relationship with the iconic and the celebrity. We love them, but then we love to see them fall over. So everybody loved Brunel's failure, and they started to make wry mocking comments about whether it would ever go into the water at all. This must have gnawed away at Brunel, because he wasn't used to this kind of failure. He was used to problems that could be solved. This was terrible. He's constantly now coming up with different ways of trying to get the ship from where she is to where she ought to be, and the country is watching anxiously. Will it move, won't it move? How many inches?'

As the short, foggy November days merged into December, a blackboard was placed by each cradle to record every movement. The men often worked into the night, their efforts illuminated by the flickering flames of open fires. However, the ship stubbornly refused to budge at anything other than a snail's pace. Brunel's problem became everybody's problem, and the letters came pouring in. A physicist suggested pulling on each cradle alternately, using small cannon to overcome the resistance. Thomas Wright from Notting Hill advised employing a troop of five hundred or more soldiers to trot to the music of a drum or fife around the deck of the ship. (The power of soldiers marching in step was well recognized, hence the signs on some older bridges warning soldiers to 'break step'.)

By mid-December, Robert Stephenson was on the scene. In the Brunel collection at Bristol University Library is a thick bundle of letters from Stephenson to Brunel at this critical time. Stephenson was not a well man, and the handwriting is large and scrawling. However, the warmth of his support shines through:

> *May I call upon you in Duke Street in the evening? Or perhaps you would rather be quiet. Don't reply to this. I will send my groom tomorrow morning for anything you may have to say at ½ past 8.*

In another note, he apologizes for having to put off a meeting on his doctor's advice. He ends with a telling P.S.:

*Never mind Russell or the papers. I shall always be at hand happen what
may to aid and do everything in my power without shirking any
responsibility if need be.*

Samuel Smiles recalled the lengths to which Stephenson would go to help his
friend. The author was with him one evening at his house in Gloucester
Square when Stephenson received a note from Brunel asking him to be at the
shipyard at six the next morning. He went as requested and stayed until
dusk. In his biography, Smiles noted:

*About midday, while superintending the launching operations, the baulk of
timber on which he [Stephenson] stood canted up, and he fell up to his
middle in the Thames mud. He was dressed as usual, without greatcoat
(though the day was bitter cold), and with only thin boots on his feet. He
was urged to leave the yard, and change his dress, or at least dry himself; but
with his usual disregard of health, he replied, 'Oh never mind me – I am
quite used to this sort of thing;' and he went paddling about in the mud,
smoking his cigar, until almost dark, when the day's work was brought to
an end. The result of this exposure was an attack of inflammation of the
lungs, which kept him to his bed for a fortnight.*

Encouraged by Stephenson's loyal support, Brunel decided to try a more
straightforward – if expensive – launching method: that of simply applying
more power. He sent his assistants round the country in search of hydraulic
presses. One firm that benefited from this desperate search was Tangye
Brothers in Birmingham. Its members later claimed: 'We launched the *Great
Eastern* and she launched us.' Other companies lent hydraulic presses free
of charge; among them was the large one that had been used for lifting the
tubes of the Britannia Bridge.

The new presses were tested, and all through a freezing January,
Stephenson was at Brunel's side, watching and advising. When at last the
moment for the final launch came, Stephenson was ill in bed. His frustration
at being so far from the action shines through in his letters – but on
1 February 1858, the note changes to one of satisfaction.

My Dear Brunel—
*I slept last night like a top, after I received your message. I got desperately anxious
all day, but my doctor would not permit me to venture so far away as Millwall.*

I do, my good friend, most sincerely congratulate you on the arrival of
the conclusion of your anxiety.
Yours sincerely,
Robert Stephenson

At daybreak that morning, the bolts had been removed from the *Great Eastern*'s cradle wedges for the last time. The hydraulic presses were started up, the winches manned, and all the men watched as an exceptionally high tide lapped against the iron side of the waiting ship. By the early afternoon, the ship was afloat. There was one tense moment when the starboard paddle wheel fouled a moored barge, but the barge was scuttled and the leviathan continued her majestic progress. By the evening, she was securely at her Deptford moorings.

The *Great Eastern* had finally taken to the waves – but at a terrible cost in mind and body not only to her creator, but also to his lifelong friend and rival.

'SEE THE CONQUERING HERO COMES'

Away from the spell of the *Great Eastern*, Brunel did his best to recover his health and spirits. On holiday in Cairo, he rode about the streets on a donkey, and met up with Robert Stephenson, who had sailed to Egypt on his yacht, *Titania*.

Brunel and his family stayed, typically, in the exotic Hotel d'Orient; Stephenson preferred that time-honoured retreat of the Englishman abroad, Shepheard's Hotel. On Christmas Day 1858, they met up for dinner. Brunel was suffering from chronic nephritis, then known as Bright's disease; Stephenson had what his biographer described as a 'deep-seated mischief in liver, stomach and nerves'. Their bodies, like the steam engines that were their stock in trade, were simply worn out.

This Christmas conversation was the last recorded meeting between the two men. They would certainly have had a lot to talk about, and no doubt managed a few convivial moments despite their various disorders. Stephenson, his health temporarily restored, went back to Britain in February 1859 and became much exercised about questions of metropolitan drainage. Brunel was not to return until May.

Meanwhile, the bridge at Saltash – now officially named the Royal Albert Bridge – had been opened to traffic. When the first truss had been successfully floated, Brunel had stood in a prominent position while a band played 'See the Conquering Hero Comes'. On 10 May 1859, the conquering hero, broken from overwork, could not even stand up. He had to lie on a couch, and be drawn across the bridge by a locomotive.

In London, the final touches were being added to the glittering interior of the *Great Eastern*. Scott Russell had managed to retain his grip on the *Great Eastern* contract, and the ship was due to sail in September for a triumphal tour of south coast ports before embarking on a trial run to America.

Brunel could not keep away, despite his increasing weakness. Most days found him on board the ship, supervising last-minute preparations. On 5 September, the day before she was due to sail, he arrived early. The London Stereoscopic Company visited to take views for the new Stereoscopic Box, a favourite drawing-room curiosity. Brunel posed for his picture by the ship's funnel. Leaning on a stick, his trademark stovepipe hat in his hand, this is a portrait of the lion tamed. Brunel was a dying man; not long afterwards, he was felled by a paralyzing stroke.

△ Crowds wave as Prince Albert prepares to board the train for the official opening ceremony of the bridge at Saltash named in his honour, May 1859.

△ Previous page: Brunel, anxious and covered in mud, photographed by Robert Howlett against the chains used to launch the *Great Eastern*.

△ The last photograph taken of Brunel on the deck of the *Great Eastern*, 5 September 1859. Within hours, he was to suffer a stroke.

Lying on his invalid's couch in Duke Street, he lingered on for a few days. His last letter, dictated and signed on 9 September, was a request that the men from the GWR engine works at Swindon should be given special passes to visit the *Great Eastern* when she reached Weymouth. However, that same day disaster struck.

The *Great Eastern* was cruising down the English Channel at speed when a terrific explosion shook the ship. Amid a cloud of steam, the huge forward funnel launched itself into the air, together with a large section of the deck and saloon. Scalding water cascaded into a boiler room, inflicting horrific injuries from which five stokers died. Such an explosion would have wrecked any other ship, but the *Great Eastern* simply steamed on.

News of the accident – which had been caused by a failure to open a stopcock – was the final blow for the ship's creator. Within six days, Brunel had died. The GWR locomotive engineer Daniel Gooch wrote this tribute in his diary:

On the 15 September 1859 I lost my oldest and best friend in the death of Mr Brunel... By his death the greatest of England's engineers was lost, the man of the greatest originality of thought and power of execution, bold in his plans, but right.

The day before Brunel's death, Robert Stephenson had arrived back in England on his yacht, *Titania*. He had been in Norway at celebrations for the opening of a new railway, but had been struck by an acute attack of jaundice. He struggled back to London only to hear that his friend was on his deathbed. Stephenson rallied for anofther month, but had resigned himself to the fact that he was next in line. He died on 12 October 1859.

Both men had worked themselves to death, but the way their country bade farewell to them could not have been more different. Robert Stephenson was buried in Westminster Abbey, next to Thomas Telford. Queen Victoria gave special permission for the cortege to pass through Hyde

Park, and 3,000 tickets were sold to spectators. Shipping lay silent, with flags at half-mast. In Tyneside, all business stopped at noon on the day of the funeral. In Newcastle, the employees of Robert Stephenson and Co marched through silent streets to a memorial service.

In contrast to this, Brunel had a relatively quiet burial in the family grave in Kensal Green Cemetery. This was the necropolis that his father, Marc, had helped Augustus Charles Pugin to design, as a hygienic solution to the disposal of London's dead.

'It's in some ways revealing that Robert Stephenson was celebrated in Westminster Abbey like a poet or a warrior, whereas Brunel was buried by his family in a new, modern cemetery plot,' says Andrew Lambert.

'Interring the dead in central London by the 1850s was thought to be very antediluvian and

△ Brunel's Hungerford Suspension Bridge, chains from which were used to build the Clifton Suspension Bridge in his memory.

rather vulgar. Brunel certainly wouldn't have wanted to be alongside Thomas Telford, for example, in a pantheon of engineers. He was a different kind of engineer and his loyalties were not ultimately there, they were with his father.'

Brunel's obituary in *The Times* set out all his faults; Stephenson's hailed him as one of the greatest engineers of all time. However, Brunel did have two memorials that showed the respect in which he was held by his fellow engineers. One was a plaque on the Royal Albert Bridge at Saltash, inscribed simply: I K BRUNEL ENGINEER 1859. The other was the Clifton Suspension Bridge, which was completed in his honour using funds raised specially by the Institution of Civil Engineers.

The completion of the Clifton Bridge was a bittersweet tribute, since the chains used to support it came from another Brunel bridge that was then in the process of being demolished. This was the Hungerford Suspension Bridge, a footbridge that had had to go to make way for the Charing Cross Railway Bridge. However, it is clear from Brunel's own writing that the Hungerford bridge was never a project he rated highly. As he wrote in his diary in 1835: 'I have condescended to be engineer to this, but I shan't give myself much trouble about it.'

'Brunel could make an engineering epic, but not an engineering sonnet'

△ Brunel's Clifton Suspension Bridge, designed in the 1830s and built in the 1860s, now carries four million cars a year. The bridge we know is a simplified version of Brunel's original design.

was how the *Morning Chronicle* summed up this lack of enthusiasm for lesser projects. It was a tendency that came with a price tag, as this newspaper's obituary clearly recognized:

Not one of the great schemes which he set on foot can fairly be called profitable, and yet they are cited, not only with pride, but with satisfaction, by the great body of a nation supposed to be pre-eminently fond of profit; and the man himself was, above all other projectors, a favourite with those very shareholders whose pockets he so unceasingly continued to empty.

There is always something not displeasing to the British temperament in a magnificent disappointment.

It is true that several of Brunel's projects during his lifetime appeared to be bottomless pits in which to sink money, resulting only in 'magnificent disappointment'. However, even his most maverick schemes often proved a stimulus for other inventors. The atmospheric railway may have been the most expensive engineering failure of its time – but pneumatic railways still exist, for example in Jakarta, Indonesia. And who knows, perhaps a search for alternative fuel sources might revive this Victorian fad.

The Thames Tunnel – nicknamed the Great Bore because it took so long to finish – was finally opened in 1843. A song was even written to celebrate its delights, and it is largely due to the pioneering techniques developed to build it that, throughout the following century, British engineers were to lead the world in tunnelling. Funding to build the road access ramps was never found, so it was only accessible by walking down spiral stairways in the construction shafts. For years, it carried only a trickle of pedestrians, who could hear the sound of ship's propellers in the river above, because of the closeness of the riverbed. Eventually, it was bought to form part of an underground rail link, and today if you take a tube between Rotherhithe and Wapping on the East London line, you will be using the Thames Tunnel.

THE BURIAL OF THE "BROAD-GAUGE."

Brunel's genius was that of the alchemist. He took other people's ideas and transformed them into something uniquely his own. When he designed his Great Western Railway, he borrowed Stephenson's plans for the London and Birmingham Railway, and used them as a model for his new line. Where Stephenson's railway looked back to the past – for instance, in his use of stationary engines to haul trains up the slope from Euston Station – Brunel's looked to the future. It is no coincidence that the Great Western Railway was the first used by the Intercity 125 high-speed train. His fanatical enthusiasm for the broad gauge might seem perverse – especially when every other railway engineer in the country was working to the standard gauge. It also cost the GWR vast amounts of money when it was eventually forced to scrap its fleet of broad-gauge engines and convert to narrow gauge. However, the competition between the two gauges undoubtedly helped to produce better locomotives. It was all part of Brunel's vision for the smoothest, fastest ride in the country – the 'Billiard Table'.

'People will wax lyrical about the GWR until the end of time,' says Andrew Lambert. 'It's still the best railway in England. Nobody will ever wax lyrical about Robert Stephenson's railway lines. London to Birmingham is not a great journey and the railway line doesn't do anything for you – it's just a way of getting from A to B, which is all it was ever meant to be.

△ *Punch*'s vision of the ghost of Brunel, lamenting the death of the broad gauge, 4 June 1892. Starting shortly after Brunel's death, the GWR gradually converted all its track to standard gauge.

▽ Following pages: Robert Stephenson's High Level Bridge over the River Tyne (1849), which has trains on the top deck and a road below.

△ Miles of extra sidings had to be laid at Swindon to accommodate redundant locomotives as they awaited scrapping or conversion. The photograph on page 66 showed the last broad-gauge through train being given a formal send-off as it left Paddington Station, on 20 May 1892. The engine would have ended up here.

'Brunel, in making a railway, makes something far bigger, far grander. He makes something that catches the imagination and opens people's eyes and gives them a new way of thinking about the world. Robert Stephenson just gets them to work on time.'

Whatever the merits of their individual engineering feats, both men altered the landscape of Britain forever. Brunel's bridge at Saltash, and Stephenson's High Level Bridge over the Tyne at Newcastle stride on giant legs above the mere humans below. Their railways altered people's mental map of the countryside, reducing it to a network of straight, flat lines.

People now complain about the disorientating effect of jetting from one airport to another. However, this is simply a logical extension of the speeding-up process that has been going on since the 19th century. To those who were used to the measured pace of stagecoach travel, the advent of the railway must have come as a severe shock. Railways annihilated space and time, even causing the clocks to be changed. (Railway time replaced local time so that standardized timetables could be introduced.) Passengers became living parcels, to be transported from A to B as quickly and directly as possible.

The politician Thomas Creevy, an early passenger on one of Robert Stephenson's locomotives, commented: 'It is really flying, and it is impossible to divest yourself of the notion of instant death to all upon the least accident

happening.' His words were proved prophetic when a fellow politician, William Huskisson, the MP for Liverpool, fell under the wheels of the *Rocket* on the opening day of the Liverpool and Manchester Railway. Huskisson survived only a few hours, becoming the first passenger – but by no means the last – to die in a railway accident.

During the gauge wars, the speed at which the GWR trains could go was cited by the pro-broad-gauge lobby as an advantage, and by the opposition as a hazard. The very qualities that made them the *Concorde* of the railways also made them the most dangerous. Frederick S Williams, in *Our Iron Roads*, published a list of accidents and 'injuries to life and limb' sustained on railways throughout Britain in the six months up to 31 December 1851. These included the following cases:

△ William Huskisson – the first fatal railway accident victim. Huskisson had supported the railway; ironically he paid for that support with his life.

> *Alexander Topping, guard, walking along the top of a train in motion (contrary to rule), came in contact with a bridge and was killed… J Reid, platelayer, run over while asleep on the Mold line… Jeremiah Fisher, said to be imbecile, stepped on the line in front of an engine and was killed… John Nicholson, driving an ass and cart through the open gates at a level crossing, turned along the line, and was run over and killed by a train…*

It is worth remembering that Huskisson died because he failed to realize that it was not a particularly bright idea to walk across railway tracks. Many of the early accidents were caused by a similar lack of understanding of the dangers of steam locomotives. From the outset, railways were built by private companies, and the need for safety was always balanced against the need to make a profit. With modern high-speed trains, the risks are greater, but the problem remains the same.

Railways were to the 19th century what computers and the Internet have been to our own era. Excited by the technology of the future, many ordinary people invested money they could ill afford to lose in a market that could not expand indefinitely. Just as fortunes were lost when the 'Dot Com' bubble collapsed in April 2000, Victorian speculators were severely burnt when

▷ The dangers of railways. One of the worst head-on collisions in British railway history occurred on 10 September 1874, at Thorpe near Norwich. Two trains had been mistakenly dispatched from either end of a single line, killing 25 people and injuring 75.

the bottom fell out of the railway share market. As Samuel Smiles put it:

> *The mania was not confined to the precincts of the Stock Exchange, but infected all ranks. It embraced merchants and manufacturers, gentry and shopkeepers, clerks in public offices, and loungers at the clubs. Noble lords were pointed to as 'stags'; there were even clergymen who were characterised as 'bulls;' and amiable ladies who had the reputation of 'bears' in the share markets. The few quiet men who remained uninfluenced by the speculation of the time were, in not a few cases, even reproached for doing injustice to their families, in declining to help themselves from the stores of wealth that were poured out on all sides.*

Figures like George Hudson, nicknamed the 'Railway King', became the bogeymen of disappointed shareholders. Thrown out of his Yorkshire village for fathering an illegitimate child when he was fifteen, Hudson started his working life in a draper's shop.

Through shrewd investment in railways, Hudson expanded his empire from Berwick on Tweed in the North down to Cambridge and Colchester in the South. At the end of 1848, there were just over 5,000 miles (8,000 km) of railways in operation in Britain, and Hudson controlled 1,500 miles (2,400 km) of them.

In his early days, Hudson had worked closely with George Stephenson – a fact conveniently forgotten by Stephenson's biographers when Hudson's name later became linked with financial sharp practices. However, unlike Stephenson, Hudson was no engineer – the only thing he made was money.

Hudson flaunted his wealth. He had two Yorkshire estates, and a five-storey Italianate mansion called Albert Gate, on the edge of London's Hyde Park. His wife was notorious for her malapropisms, bad dress sense and social gaffes. (The story goes that, when asked whether she would like sherry or port, she replied, 'A little of both, please'.) No doubt it was partly snobbery that led *Punch* to dance on the grave of Hudson's reputation after his empire came crashing down. However, his lax accounting and dubious dealings cost many people dearly. Railway share speculation crossed social boundaries; a famous *Punch* cartoon showed an anxious Queen Victoria posing the question to her beloved consort: 'Tell me, Oh, tell me, dear Albert. Have you any railway shares?'

It was Brunel's disgust with Railway Mania, mixed with his usual restlessness, that was partly responsible for his change of focus from railways to ships. However, the fact that he relished the challenge underlines the fact that he was more of a polymath than Stephenson.

In 1857, the *Great Western* made her last sad journey up the River Thames to the breaker's yard. Brunel, then heavily engaged with the *Great Eastern*, still found time to pay a farewell visit to her berth at Vauxhall, before her timbers were finally pulled apart. Pictures, written records and the ship's bell are all that now remain of her. The bell is on display at the Great Western Dock, Bristol, near the *Great Britain* – a ship that has had a very different history.

△ George Hudson was a favourite target for satirists, as in this *Punch* cartoon, entitled 'Off the Rails'. Charles Dickens wrote: 'There are some dogs who cannot endure one particular note on the Piano. In like manner, I feel disposed to throw up my head, and howl, whenever I hear Mr Hudson mentioned.'

For a quarter of a century, the *Great Britain* carried emigrants to and from Australia. She carried troops and horses to the Crimea, and to Bombay at the time of the Indian Mutiny. During her last incarnation as a cargo ship she was damaged by a storm while rounding Cape Horn, and landed up in the harbour of Port Stanley, in the Falkland Islands. For years, she was used first to store wool, then coal. However, she started to leak, and in April 1937 was towed 3 miles (5 km) out of the harbour to be scuttled at Sparrow Cove, the haunt of rare sea birds but not much else. Holes were blasted in her sides, and she was left to her fate.

Over the years, souvenir hunters stripped her of much that was portable, but in the 1960s, a growing interest in conservation revived interest in the ship. The marine architect, Ewan Corlett, wrote a letter to *The Times*, which sparked a massive fund-raising campaign. The ship was patched up, and floated on to a giant pontoon. She was lashed down to this pontoon for the slow 7,000-mile (11,263-km) tow home to Bristol, one of the longest ever attempted in maritime history.

Having safely made the Atlantic crossing, she was watched by a crowd of over 100,000 people as she finally made her way up the river to the Bristol Docks. On her first passage along the Avon, on 11 December 1844, the Clifton Suspension Bridge had not been completed. Now, 126 years later, she passed underneath its lofty roadway. The moment was a double reminder of how much Brunel's reputation had grown since his death.

Andrew Lambert describes the iron *Great Britain* as the first modern ship; in comparison, the wooden *Great Western* was the last of the old ones. As he puts it:

'A wooden ship is a work of art. No two wooden ships are ever the same – they have to be fashioned out of a completely unreliable material, essentially by hand. An iron ship is made out of a reliable material and you can use the same processes over and over again.

'You can build iron ships in exactly the same way that you can build motor cars – you can build them all the same size and shape. In the Second World War the Americans were building hundreds of ships to a standard design. That's what you get – reliability, standardisation, measurement, precision.'

When Brunel first conceived of his most ambitious ship of all, the *Great Eastern*, he originally planned to build two at once, both exactly the same. However, it soon became obvious that this would be impractical; as things turned out, it was just as well he had only one ship to launch.

The size and shape of the *Great Eastern* caused problems long after the launch was successfully accomplished. Few ports could offer quays with water deep enough for her, and her shape turned out to produce an unpleasant rolling motion in mid-ocean. In the short term, this made her passengers seasick; however, in the long term, it produced data that prevented future ocean liners suffering the same problem. Shipbuilders also learnt from her unique construction. The idea of a cellular double bottom, which gave her such strength, was adopted by other large ships, and is now being made compulsory for supertankers.

As a passenger ship, the *Great Eastern* was never filled to capacity. Brunel had simply misjudged the market. However, she took on a new lease of life when she was chartered by the Telegraph Construction Company to lay the first communication cable under the Atlantic. A prime mover in this effort was Daniel Gooch, Brunel's old GWR friend and colleague.

Cabins were taken out and replaced by cable tanks, and the *Great Eastern* started on her new duties in 1865. Her first attempt failed, when the cable broke and sunk in more than 2 miles (3 km) of water. However, the second attempt the following year was successful. On Thursday 26 July 1866, Daniel Gooch wrote in his diary:

△ The *Great Eastern* became the first ship to lay a permanent transatlantic communications cable. Here, workers search for faults after recovering the cable from the bed of the Atlantic in July 1866.

We have achieved our great object and laid our cable from shore to shore, along which the lightning may now flash messages of peace and goodwill between two kindred nations. Is it wrong that I should have felt as though my heart would burst when that end of our long line touched the shore amid the booming of cannon, the wild half-mad cheers and shouts of the men? It seemed more than I could bear.

One of the first messages to be sent in the dots and dashes of Morse code was to the Foreign Office in Britain, announcing the successful link. A reply came the next morning, in the form of congratulations from Queen Victoria to the President of the United States of America.

▽ The *Queen Mary 2* under construction in April 2003. The largest, longest, tallest, most expensive passenger vessel ever built, this is a ship in the *Great Eastern* tradition.

After a successful career laying cables all around the world, the *Great Eastern* was converted back to a passenger ship, but did not prosper. For a while, she suffered the ignominy of being turned into a showboat at Liverpool, before being broken up. Some have presented her simply as an example of Brunel's megalomania. However, a large part of the *Great Eastern's* problem was that she was an idea whose time had not yet come. Today's giant liners, such as the £538 million *Queen Mary 2*, launched from St Nazaire Dock in Brittany in 2003, owe a direct debt to Brunel's vision.

Our risk-management culture is only prepared to tolerate success – yet much of the knowledge base of today's engineers has been gained by learning from others' mistakes. As Simon Schaffer points out, our predecessors were prepared for catastrophic failure in a way that we are not:

'It's not just that mid-Victorian engineers were risk-takers, although they certainly were,' he says. 'It's that catastrophe was treated as part of a kind of evangelical experience. For a culture dominated by evangelical Christianity, it was understood that humans were put on the earth to suffer and die. This was not the age of achievement, this was the age of atonement, and suffering and failure was a sign of one's Christ-like patience in the face of appalling experiences.

'If you apply that to the world of the engineers, what that meant was that they were constantly pushing at the envelope of the feasible, they were

△ Painting the Saltash Bridge in the 1930s – a risky business in itself.

constantly trying to see whether there was a subtly better way of engineering a particular solution.

'This is an engineering culture that massively learns from experience but is never satisfied with it. Each new bridge or each new railway, each new dam or each new tunnel, is understood as being an experiment based on previous experience. What it's not is a simple replication of that experience.

'One sees in Brunel's projects the way in which successive bridge designs are a series of developing experiments. If one compares, say, the use of metal in the Chepstow Bridge with that at the triumphant bridge at Saltash, Chepstow is an experiment for what is going to be done at Saltash. The idea is not that a bridge should be a completed project – it's an indication of what might happen next.'

In 2003, the Royal Albert Bridge at Saltash came up for renovation. A firm of engineering consultants was brought in to try and reconstruct what Brunel was thinking when he built the bridge. Only after extensive research and testing were they able to work out the full subtlety of the design.

Chris Wise, Professor of Civil Engineering Design at Imperial College, London, has admired the Saltash bridge ever since he first saw it as a child:

'It's a bit of a mongrel, actually – there are three different sorts of bridges hidden in here,' he says. 'You might say that's slightly over the top, but the very over-the-topness of it means it works really, really well. There isn't anything exactly like this around the world, as far as I know.'

Just for his own interest, he built a model of the bridge in his computer, and sent a virtual train across it:

'If it's not a very good design, what happens is the bridge has a wave going across it. As the train comes across, the wave goes in front of it. You can imagine, that tortures the metal and eventually it just fails. What I found out on this bridge is that the very light cross bracing that Brunel dismisses and says "actually it wasn't doing much" is very important. As the railway train goes across, it makes the bridge behave a little bit like a big beam – very strong. So strong, in fact, that when the train goes across, the bridge only moves by half an inch.

'Having done that, I thought I'll try and beat Brunel. I'll see if I can come up with something better than him. So I tried five different variations on this bridge to see if I could improve on it. I had a sneaking feeling that I wouldn't get it any better, but every single one I did actually made it worse. I managed to get to the point where, if you turned it into a suspension bridge, it would have fallen down the first time a train went across it.'

As an engineer who often works on high-profile projects, Chris Wise finds it very easy to empathize with Brunel – especially at his most public moments, such as the floating of the Saltash trusses or the launching of the *Great Eastern*. As he puts it: 'Engineering's a bit like sport, because something might go wrong. You never know until the final whistle blows what's going to happen.'

His own most embarrassing moment came on 10 June 2000. The scene – as with Brunel's *Great Eastern* – was the River Thames. The occasion was the opening of central London's first new river crossing for more than a century. The £18.2 million Millennium Bridge was a collaboration between Norman Foster's architectural firm, the British sculptor, Sir Anthony Caro, and the consulting engineers Arup, of which Chris Wise was a director. As he recalls:

'When we opened the Millennium Bridge, there were people queued up ten-deep on the banks of the River Thames, both sides, for as far as you could see up and down. This is in a part of London when on a Saturday or a Sunday there's normally nobody.

'At least we hadn't asked people to pay for the privilege of watching it – I would have felt particularly embarrassed about that. But I did feel like Brunel did when the *Great Eastern* jammed on the slipway.

'I was standing there with my nine-year-old son and thousands of people queuing up on the banks of the river wanted to get onto the bridge. They started walking across the bridge and after thirty seconds it started moving. By that time there were about a thousand people on the bridge and I had a moment where I really did think, "Crikey! We're all going to end up in the river here."'

After a short while, the vibration settled into a pattern and did not get any worse.

'At that point – and I'm sure Brunel would have found the same thing – I wasn't worried about the consequences any more. I wanted to know what had gone wrong. I found myself looking at my watch, and I looked around and all the engineering mates I had were also looking at their watches. They were trying to time the vibrations to see what it was that was causing it. They were trying to understand it.'

The team working on the Millennium Bridge were well aware that it was a cutting-edge project, and there was a small chance it might go wrong. However, it is no longer in the culture to admit this possibility. That is why the 'wobbly bridge' hit the headlines.

△ The £18.2 million Millennium Bridge, nicknamed the wobbly bridge. It was closed for almost a year and a half while engineers worked to stabilise it.

Today, the engineer is often simply the person brought in to make sure a particular architect's imaginings remain standing. Engineering has become a highly specialized activity – a process that began in the mid-Victorian era with the formation of separate institutes for civil and mechanical engineers. Brunel and Stephenson played the roles of engineer, architect, fund-raiser and public relations officer all in one. With his great ships, Brunel also put his hand in his own pocket – unlike Stephenson, who rarely even took shares in his own projects.

They were heroes in their lifetime, a process that was continued by their biographers after their death. Brunel's son, Isambard, concluded his account of his father's work with an address given by the railway engineer Joseph Locke to the Institution of Civil Engineers. This was a tribute to both Brunel and Stephenson; however, it included an important caveat:

Man is not perfect, and it is not to be expected that he should be always
successful; and, as in the midst of success we sometimes learn great truths

before unknown to us, so also we often discover in failure the causes which frustrate our best-directed efforts.

Robert Stephenson's biographers, J C Jeaffreson, and Samuel Smiles, glossed over failures, or at least presented them in the brightest colours. Smiles was the great Victorian hagiographer, and wherever possible he fitted his subjects into the mould of the modest, self-made man. George Stephenson was perfect from this point of view, as he had started from such humble beginnings. Robert was more problematic, however, since he had obviously benefited from his father's success. Perhaps for this reason, Smiles went out of his way to stress Robert's modesty. It is largely due to his account of Robert's life, presented as an appendix to the biography of his father in the *Lives of the Engineers* series, that history has credited George Stephenson with many achievements that rightfully belong to his son Robert.

Smiles wrote about Brunel briefly in a review of Beamish's life of Marc Brunel. However, Isambard did not fit comfortably into the saintly mould; he could hardly be said to have been either modest, or a self-made man. It was easy for journalists to turn Robert Stephenson into a moral example; it was almost impossible to do the same for Brunel.

Both men declined knighthoods during their lifetime, but both accepted honorary degrees from the University of Oxford. Their only real interest was in recognition by their peers, and this they achieved in abundance. However, their iron dreams would have come to nothing without the efforts of thousands of nameless navvies who toiled in their service. As Elizabeth Garnett, secretary to the Navvy Mission Society, wrote in 1879:

> *Certainly no men in all the world so improve their country as Navvies do England. Their work will last for ages, and if the world remains so long, people will come hundreds of years hence to look at and to wonder at what they have done.*

For Brunel and Stephenson, the best epitaph would be the lines written for Sir Christopher Wren over the interior of the North Door in St Paul's Cathedral, London: *'Si monumentum requiris, circumspice.'* If you would see their monument, look around.

▷ The following places have surviving examples of engineering or are mentioned in the text.

1: Caledonian Canal
– Thomas Telford

2: Balmoral
• Bridge across the River Dee – Brunel

3: Newcastle
• High Level Bridge over the Tyne – Robert Stephenson
• Stephenson Railway Museum

4: York
• National Railway Museum

5: Menai Straits
• Britannia Bridge – Robert Stephenson
• Menai Straits Bridge – Thomas Telford

6: Conway Bridge
Robert Stephenson

7: Shropshire Union Canal
• Pontcysyllte Aqueduct – Thomas Telford

8: St Helens
• Sankey Viaduct

9: Saltash
• Royal Albert Bridge – Brunel

10: Starcross
• Atmospheric Pumping Station

11: Bristol
• Clifton Suspension Bridge – Brunel
• SS *Great Britain* – Brunel
• Temple Meads Station – Brunel

12: Box
• Tunnel – Brunel

13: Swindon
• Museum of the Great Western Railway
• GWR Railway Village

14: Didcot
• Didcot Railway Centre

15: Culham
• Railway Station – Brunel

16: Maidenhead
• GWR bridge – Brunel

17: Hanwell
• Wharncliffe Viaduct – Brunel

18: London
• Brunel Engine House, Rotherhithe
• Kensal Green Cemetery
• Paddington Station
• Science Museum
• National Maritime Museum, Greenwich
• Westminster Abbey

ORIGINAL SOURCES

Lady Celia Noble, Brunel's granddaughter, left her extensive collection of Brunel's letters, diaries and journals to the University of Bristol. Original sources quoted in the text come from the university's Brunel Collection, unless otherwise stated.

GENERAL

Brunel, Isambard, *The Life of Isambard Kingdom Brunel, Civil Engineer.* Longmans, London, 1870

Buchanan, Angus, *The Life and Times of Isambard Kingdom Brunel.* Hambledon, London, 2002

Jeaffreson, J C, *The Life of Robert Stephenson.* Two vols, Longmans, London, 1864

Kentley, Eric (Ed), *Isambard Kingdom Brunel, Recent Works.* Design Museum, London, 2000

Noble, Celia Brunel, *The Brunels, Father and Son.* Cobden-Sanderson, London, 1938

Rolt, L T C, *George and Robert Stephenson, The Railway Revolution.* Longmans, London, 1962

Rolt, L T C, *Isambard Kingdom Brunel.* Longmans, London, 1957; Penguin paperback 1970

Vaughan, Adrian, *Isambard Kingdom Brunel: Engineering Knight-Errant.* John Murray, London, 1991

CHAPTER ONE

Beamish, Richard, *Memoir of the Life of Sir Marc Isambard Brunel.* Longmans, London, 1862

Chrimes, Michael and others, *The Triumphant Bore: A Celebration of Marc Brunel's Thames Tunnel.* Thomas Telford, London, 1993

CHAPTER TWO

Body, Geoffrey, *Clifton Suspension Bridge, An Illustrated History*. Moonraker, Bradford-on-Avon, 1976

Burton, Anthony, *Thomas Telford*. Aurum, London, 1999

Kemble, Frances Anne (Fanny), *Records of a Girlhood*. Two vols, Bentley, London, 1878

Smiles, Samuel, *The Life of George Stephenson*. John Murray, London, 1873

Smiles, Samuel, *The Life of Thomas Telford, Civil Engineer*. John Murray, London, 1867

Thomas, R H G, *The Liverpool and Manchester Railway*. Batsford, London, 1980

CHAPTER THREE

Hyman, Anthony, *Charles Babbage: Pioneer of the Computer*. OUP, Oxford, 1982

MacDermot, E T, *History of the Great Western Railway Vol 1, Revised by Clinker, C R*, London, 1964

Simmons, Jack (Ed), *The Men Who Built Railways*. A reprint of F R Conder's *Personal Recollections of English Engineers*. Thomas Telford, London, 1983

Simmons, Jack, *The Victorian Railway*. Thames & Hudson, London, 1991

CHAPTER FOUR

Ball, Adrian, and Wright, Diana, *SS Great Britain*. David & Charles, Newton Abbot, 1981

Griffiths, Denis, *Brunel's Great Western*. Patrick Stephens, Wellingborough, 1985

Griffiths, Denis, Lambert, Andrew, and Walker, Fred, *Brunel's Ships*. Chatham, London, 1999

Smiles, Samuel (Ed), *James Nasmyth, Engineer. An Autobiography*. John Murray, London 1883

CHAPTER FIVE

Hadfield, Charles, *Atmospheric Railways. A Victorian Venture in Silent Speed*. David and Charles, Newton Abbot 1967

Wilson, R B (Ed) *Sir Daniel Gooch: Memoirs and Diary*. David & Charles, Newton Abbot, 1972

CHAPTER SIX

Brooke, David, *The Railway Navvy, 'That Despicable Race of Men'*. David & Charles, Newton Abbot, 1983

Coleman, Terry, *The Railway Navvies. A History of the Men who Made Railways.* Hutchinson, London, 1965

Tregelles, Anna, *The Ways of the Line.* William Oliphant & Co, Edinburgh, 1847

Williams, Frederick S, *Our Iron Roads.* Ingram, Cooke & Co, London, 1852

CHAPTER SEVEN

Binding, John, *Brunel's Royal Albert Bridge.* Twelveheads Press, Truro, 1997

CHAPTER EIGHT

Griffiths, Denis, '*The Steamship Great Eastern*' in *Brunel's Ships.* Chatham, London, 1999

Hobhouse, Christopher, *1851 and the Crystal Palace.* John Murray, London, 1950

Richardson, Harriet (Ed), *English Hospitals 1660–1948. A Survey of their Architecture and Design.* Royal Commission on the Historical Monuments of England, 1998

CHAPTER NINE

Beaumont, Robert, *The Railway King: A Biography of George Hudson.* Hodder Headline, London, 2002

Corlett, Ewan, *The Iron Ship*: *The History and Significance of Brunel's 'Great Britain'.* Moonraker, Bradford-on-Avon, 1975

Schivelbusch, Wolfgang, *The Railway Journey: The Industrialization of Time and Space in the 19th Century.* Berg, Leamington Spa, 1986

INDEX

PICTURE CREDITS